Springer Monographs in Mathematics

Igor Chudinovich   Christian Constanda

Variational and Potential Methods for a Class of Linear Hyperbolic Evolutionary Processes

 Springer

Igor Chudinovich, MS, PhD, DSc
Professor of Mathematics, Faculty of Mechanical, Electrical, and Electronic Engineering
The University of Guanajuato, Salamanca, GTO, Mexico

Christian Constanda, MS, PhD, DSc
Charles W. Oliphant Professor of Mathematical Sciences, The University of Tulsa
600 South College Avenue, Tulsa, Oklahoma 74104, USA

British Library Cataloguing in Publication Data
Chudinovich, Igor
    Variational and potential methods for a class of linear
    hyperbolic evolutionary processes. – (Springer monographs
    in mathematics)
    1. Plates (Engineering) – Mathematical models   2. Boundary
    element methods   3. Differential equations, Hyperbolic
    4. Differential equations, Linear
    I. Title   II. Constanda, C. (Christian)
    515.3'535

Library of Congress Cataloging-in-Publication Data
CIP data available.

Mathematics Subject Classification (2000): 35C15; 35D05; 35E05; 35L15; 35L20; 35Q72; 45F15; 74H20; 74H25; 74K20

Springer Monographs in Mathematics ISSN 1439-7382
ISBN 978-1-84996-946-8   e-ISBN 978-1-84628-120-4
Springer is a part of Springer Science+Business Media
springeronline.com

12/3830-543210   Printer on acid-free paper

For Olga and Lia
and the younger generation
Genia and Dan

Preface

Variational and boundary integral equation techniques are two of the most useful methods for solving time-dependent problems described by systems of equations of the form

$$\frac{\partial^2 u}{\partial t^2} = Au,$$

where $u = u(x, t)$ is a vector-valued function, x is a point in a domain in $\mathbb{R}^2$ or $\mathbb{R}^3$, and A is a linear elliptic differential operator. To facilitate a better understanding of these two types of methods, below we propose to illustrate their mechanisms in action on a specific mathematical model rather than in a more impersonal abstract setting. For this purpose, we have chosen the hyperbolic system of partial differential equations governing the nonstationary bending of elastic plates with transverse shear deformation. The reason for our choice is twofold. On the one hand, in a certain sense this is a "hybrid" system, consisting of three equations for three unknown functions in only two independent variables, which makes it more unusual—and thereby more interesting to the analyst—than other systems arising in solid mechanics. On the other hand, this particular plate model has received very little attention compared to the so-called classical one, based on Kirchhoff's simplifying hypotheses, although, as acknowledged by practitioners, it represents a substantial refinement of the latter and therefore needs a rigorous discussion of the existence, uniqueness, and continuous dependence of its solution on the data before any construction of numerical approximation algorithms can be contemplated.

The first part of our analysis is conducted by means of a procedure that is close in both nature and detail to a variational method, and which, for this reason, we also call variational. Once the results have been established in the general setting of Sobolev spaces, we carry out the second part of the study by seeking useful, closed-form integral representations of the solutions in terms of dynamic (retarded) plate potentials.

The problems discussed in this book include those with Dirichlet and Neumann boundary conditions (corresponding, in particular, to the clamped-edge and free-edge plate), with elastic (Robin), mixed, and combined displacement-traction (simply supported edge) boundary data, transmission (contact) problems, problems for plates with homogeneous inclusions, plates with cracks, and plates on a generalized elastic foundation. For each of them, the variational version is formulated and its solvability is examined in spaces of distributions; subsequently, the solutions are found in the form of time-dependent single-layer and double-layer potentials with distributional densities that satisfy nonstationary integral equations. The analysis technique consists in using the Laplace transformation to reduce the original problems to boundary value problems depending on the transformation parameter, and on establishing estimates for the solutions of the latter that allow conclusions to be drawn about the existence and properties of the solutions to the given initial-boundary value problems. The transformed problems are solved by means of specially constructed algebras of singular integral operators defined by the boundary values of the transformed potentials.

The distributional setting has the advantage over the classical one in that it enables the method to be applied in less smooth domains—for example, in regions with corners and cuts. Furthermore, Sobolev-type norms are particularly useful in the global error analysis of numerical schemes, but such analysis falls outside the scope of this book and we do not pursue it.

To the authors' knowledge, this is the first time that so many typical initial-boundary value problems have been considered in the same book for a model in conjunction with both variational and boundary integral equation methods. The text provides full details of the proofs and is aimed at researchers interested in the use of applied analysis as a tool for investigating mathematical models in mechanics. The presentation assumes no specialized knowledge beyond a basic understanding of functional analysis and Sobolev spaces.

We want to emphasize that the book does not intend to explain the mechanical background of plate theory. Details of that nature and a fuller discussion of the limitations of the model that we have chosen as our object of study can be found in the article

> J.R. Cho and J.T. Oden, A priori modeling error estimate of hierarchical models for elasticity problems for plate and shell-like structures, *Math. Comput. Modelling* **23** (1996), 117–133.

Ours is a purely mathematical that aims to acquaint the interested reader with two of the most powerful and general techniques of solution for this type of linear problem. We reiterate that the theory of bending of plates with transverse shear deformation has been selected merely as an application vehicle because of its unusual features and lack of previous strict mathematical treatment. The book is a natural complement to our earlier monograph [7], where we investigated the corresponding equilibrium problems.

Some of the results discussed below have already been announced in concise form in the literature (see [4]–[6] and [8]).

The authors would like to acknowledge help and support received from various quarters during the preparation of this book. I.C. wishes to thank his former colleagues in the Mathematical Physics and Computational Mathematics section of the Department of Mathematics and Mechanics at Kharkov National University, and his current colleagues in the Department of Mechanical, Electrical, and Electronic Engineering of the University of Guanajuato in Salamanca—in particular, Drs. Igor Chueshov, Arturo Lara Lopez, and René Jaime Rivas, for playing an instrumental role in arranging his move to Mexico. C.C. wishes to thank Dr. Bill Coberly and his other colleagues at the University of Tulsa for a departmental atmosphere that has proved highly conducive to the writing of mathematical books.

Last but by no means least, we would like to place on record the debt of gratitude that we owe our wives, *sine quibus non,* who have guided us wisely, patiently, and selflessly, by word and by deed, to exciting and challenging new shores.

Igor Chudinovich

Professor of Mathematics
University of Guanajuato
Salamanca, Mexico

Christian Constanda

C.W. Oliphant Professor
   of Mathematical Sciences
University of Tulsa, USA

April 2004

Contents

Formulation of the Problems and Their Nonstationary Boundary Integral Equations

1.1 The Initial-Boundary Value Problems

All problem statements in this chapter are formal; rigorous versions will be presented after the introduction of the necessary function spaces.

Below we consider initial-boundary value problems for the time-dependent homogeneous equations of the model with homogeneous initial data. In Chapter 9, we indicate how the general case can be reduced to the homogeneous one.

By an *elastic plate* we understand an elastic body that occupies a region $\bar{S} \times [-h_0/2, h_0/2]$ in $\mathbb{R}^3$, where S is a domain in $\mathbb{R}^2$ bounded by a simple closed curve ∂S and $0 < h_0 = \text{const} \ll \text{diam}\, S$ is called the *thickness*.

Throughout the book we use the following notation and conventions.

Unless otherwise specified, Greek and Latin subscripts and superscripts in all formulas take the values $1, 2$ and $1, 2, 3$, respectively, and summation over repeated indices is adopted.

The standard inner product in $\mathbb{R}^3$ is $(a, b) = a_i b_i$.

A generic point in $\mathbb{R}^2$ referred to a Cartesian system of coordinates in the middle plane $x_3 = 0$ of the plate is written as $x = (x_1, x_2)$.

$X = (x, t)$, where t is the time variable.

Partial derivatives are denoted by $\partial_\alpha = \partial/\partial x_\alpha$ and $\partial_t = \partial/\partial t$.

A superscript T denotes matrix transposition. A superscript $*$ denotes conjugation and transposition of a complex matrix.

The columns of a matrix M are denoted by $M^{(i)}$.

Both matrix-valued functions and scalar functions are simply referred to as functions. If $\mathcal{Y}$ is a space of scalar functions and g is a matrix-valued function, then $g \in \mathcal{Y}$ means that each entry of g belongs to $\mathcal{Y}$.

A three-component vector $q = (q_1, q_2, q_3)^\mathrm{T}$ may be written alternatively as $q = (\bar{q}^\mathrm{T}, q_3)^\mathrm{T}$, where $\bar{q} = (q_1, q_2)^\mathrm{T}$.

S^+ is the finite domain enclosed by ∂S, and $S^- = \mathbb{R}^2 \setminus (S^+ \cup \partial S)$.

The boundary ∂S is a C^2-curve with a uniquely defined outward (with respect to S^+) normal $n = (n_1, n_2)^\mathrm{T}$.

We write

$$G = S \times (0, \infty), \quad G^\pm = S^\pm \times (0, \infty), \quad \Gamma = \partial S \times (0, \infty).$$

If φ is a smooth function defined in S^+ (S^-), then φ^+ (φ^-) denotes the limiting value (if it exists) of φ as its argument tends to ∂S from within S^+ (S^-). If φ is not smooth but has a trace on ∂S, then the latter is denoted by $\gamma^+ \varphi$ $(\gamma^- \varphi)$. Since there is no danger of ambiguity, the notation remains the same for functions defined in G^+ (G^-) and their limiting values (traces) on the boundary Γ.

The operators of restriction from $\mathbb{R}^2$ (or $S^+ \cup S^-$) to $S^\pm$, or from $\mathbb{R}^2 \times (0, \infty)$ (or $G^+ \cup G^-$) to $G^\pm$, are denoted by $\pi^\pm$.

Operators of extension from ∂S to $S^\pm$, or from Γ to $G^\pm$, are denoted by $l^\pm$, respectively.

Δ is the Laplacian and δ_{ij} is the Kronecker delta.

$\mathcal{L}$ and $\mathcal{L}^{-1}$ are, respectively, the Laplace transformation with respect to t, and its inverse. The Laplace transform of a function $u(x, t)$ is denoted by $\hat{u}(x, p)$, where p is the transformation parameter.

Other notation will be introduced as the need arises.

Suppose that the material is homogeneous and isotropic, of density ρ and Lamé constants λ and μ, which satisfy the inequalities [9]

$$\lambda + \mu > 0, \quad \mu > 0, \quad \rho > 0. \tag{1.1}$$

If we denote by t_{ij}, ε_{ij}, v_i, and f_i, respectively, the components of the stress tensor, deformation tensor, displacement vector, and body force vector, then the behavior of the plate as a three-dimensional elastic body under prescribed initial and boundary conditions is governed by three main groups of equations, namely (see [14] and [17]),

the kinematic formulas

$$\varepsilon_{ij} = \tfrac{1}{2}(\partial_i v_j + \partial_j v_i); \tag{1.2}$$

the stress-strain relations (generalized Hooke's law)

$$t_{ij} = \lambda \varepsilon_{kk} \delta_{ij} + 2\mu \varepsilon_{ij}; \tag{1.3}$$

the equations of motion

$$\partial_j t_{ij} + f_i = \rho \partial_t^2 v_i. \tag{1.4}$$

In addition,

$$t_i = t_{ij} n_j$$

are the components of the stress vector on ∂S.

The model of bending of plates with transverse shear deformation that we intend to study here postulates a displacement field of the form

$$v_\alpha(x, x_3, t) = x_3 u_\alpha(X),$$
$$v_3(x, x_3, t) = u_3(X).$$

(1.5)

This assumption is valid only for plates whose ratio of thickness to diameter falls within a certain range (see the Preface).

Expressions (1.5) and the geometry of the plate suggest a way of simplifying equations (1.2)–(1.4). This is done by means of a well-known procedure that involves the use of the averaging operators $\mathcal{I}_\alpha$ and $\mathcal{J}_\alpha$, $\alpha = 0, 1$, defined by

$$(\mathcal{I}_\alpha g)(X) = h_0^{-1} \left[x_3^\alpha g(x, x_3, t) \right]_{x_3 = -h_0/2}^{x_3 = h_0/2},$$

$$(\mathcal{J}_\alpha g)(X) = h_0^{-1} \int_{-h_0/2}^{h_0/2} x_3^\alpha g(x, x_3, t)\, dx_3.$$

Specifically, setting

$$N_{\alpha\beta} = \mathcal{J}_1 t_{\alpha\beta},$$
$$N_{3\alpha} = \mathcal{J}_0 t_{3\alpha},$$
$$q_\alpha = \mathcal{J}_1 f_\alpha + \mathcal{I}_1 t_{\alpha 3},$$
$$q_3 = \mathcal{J}_0 f_3 + \mathcal{I}_0 t_{33},$$
$$h^2 = h_0^2/12,$$

system (1.4) yields the plate equations of motion

$$\partial_\beta N_{\alpha\beta} - N_{3\alpha} + q_\alpha = \rho h^2 \partial_t^2 u_\alpha,$$
$$\partial_\alpha N_{3\alpha} + q_3 = \rho \partial_t^2 u_3.$$

(1.6)

Also, from (1.2), (1.3), and (1.5), we obtain the plate constitutive relations

$$N_{\alpha\beta} = h^2 \left[\lambda(\partial_\gamma u_\gamma)\delta_{\alpha\beta} + \mu(\partial_\alpha u_\beta + \partial_\beta u_\alpha) \right],$$
$$N_{3\alpha} = \mu(\partial_\alpha u_3 + u_\alpha).$$

(1.7)

Finally, substituting (1.7) into (1.6) leads to the alternative equations of motion

$$B(\partial_t^2 u)(X) + (Au)(X) = q(X), \quad X \in G^+ \text{ or } X \in G^-,$$

(1.8)

where

$$B = \mathrm{diag}\{\rho h^2, \rho h^2, \rho\},$$

A is the matrix differential operator with entries [9]

$$A_{\alpha\alpha} = -h^2\mu\Delta - h^2(\lambda + \mu)\partial_\alpha^2 - \mu \quad (\alpha \text{ not summed}),$$
$$A_{33} = -\mu\Delta,$$
$$A_{12} = A_{21} = -h^2(\lambda + \mu)\partial_1\partial_2,$$
$$A_{\alpha3} = -A_{3\alpha} = \mu\partial_\alpha,$$

and

$$u = (\bar{u}^\mathrm{T}, u_3)^\mathrm{T},$$
$$q = (\bar{q}^\mathrm{T}, q_3)^\mathrm{T}.$$

It is easily verified that, under conditions (1.1), A is a strongly elliptic operator and satisfies Gårding's inequality [18].

The quantities $N_{\alpha\beta}$ and $N_{\alpha3}$ are the averages across the thickness of the plate of the bending and twisting moments with respect to the middle plane $x_3 = 0$, and of the transverse shear forces [9]; q_α and q_3 are combinations of the body moments and forces and of the moments and forces acting on the faces $x_3 = \pm h_0/2$.

Similarly, setting

$$N_\alpha = \mathcal{J}_1 t_\alpha,$$
$$N_3 = \mathcal{J}_0 t_3,$$

we obtain

$$N_1 = h^2\big[(\lambda\partial_\alpha u_\alpha + 2\mu\partial_1 u_1)n_1 + \mu(\partial_1 u_2 + \partial_2 u_1)n_2\big],$$
$$N_2 = h^2\big[\mu(\partial_1 u_2 + \partial_2 u_1)n_1 + (\lambda\partial_\alpha u_\alpha + 2\mu\partial_2 u_2)n_2\big],$$
$$N_3 = \mu(\partial_\alpha u_3 + u_\alpha)n_\alpha,$$

which can be written as

$$N_i = (Tu)_i,$$

where T is the matrix boundary operator with entries

$$T_{11} = h^2\big[(\lambda + 2\mu)n_1\partial_1 + \mu n_2\partial_2\big],$$
$$T_{22} = h^2\big[(\lambda + 2\mu)n_2\partial_2 + \mu n_1\partial_1\big],$$
$$T_{33} = \mu n_\alpha\partial_\alpha,$$
$$T_{12} = h^2(\lambda n_1\partial_2 + \mu n_2\partial_1),$$
$$T_{21} = h^2(\mu n_1\partial_2 + \lambda n_2\partial_1),$$
$$T_{3\alpha} = \mu n_\alpha,$$
$$T_{\alpha3} = 0.$$

From what has been said above, it is obvious that Tu is the vector of the averaged moments and shear force acting on the lateral part $\partial S \times (-h_0/2, h_0/2)$

of the boundary. The vector u is referred to as the displacement vector since it characterizes the latter uniquely in terms of the assumption (1.5).

In Chapters 2–8, we deal almost exclusively with the homogeneous equation (1.8), that is,

$$B(\partial_t^2 u)(X) + (Au)(X) = 0, \quad X \in G^+ \text{ or } X \in G^-. \tag{1.9}$$

To (1.9) we adjoin appropriate boundary conditions and homogeneous initial conditions. The functions occurring on the right-hand side in all the boundary conditions below are prescribed.

The symbolic name of each problem that we consider starts with a "D" to indicate that it is a dynamic problem. The remaining letters are fairly obvious initials related to the problem type and/or boundary condition type.

Thus, the classical interior and exterior problems (DD$^\pm$) with Dirichlet boundary conditions consist, respectively, in finding functions $u \in C^2(G^\pm) \cap C^1(\bar{G}^\pm)$ such that

$$B(\partial_t^2 u)(X) + (Au)(X) = 0, \quad X \in G^\pm,$$
$$u(x, 0+) = (\partial_t u)(x, 0+) = 0, \quad x \in S^\pm,$$
$$u^\pm(X) = f(X), \quad X \in \Gamma.$$

In the interior and exterior initial boundary-value problems (DN$^\pm$) with Neumann boundary conditions, we seek solutions $u \in C^2(G^\pm) \cap C^1(\bar{G}^\pm)$ of

$$B(\partial_t^2 u)(X) + (Au)(X) = 0, \quad X \in G^\pm,$$
$$u(x, 0+) = (\partial_t u)(x, 0+) = 0, \quad x \in S^\pm,$$
$$(Tu)^\pm(X) = g(X), \quad X \in \Gamma.$$

Consider two open arcs ∂S_1 and ∂S_2 of ∂S such that

$$\mathrm{mes}(\partial S_\alpha) > 0,$$
$$\overline{\partial S_1} \cup \overline{\partial S_2} = \partial S,$$
$$\partial S_1 \cap \partial S_2 = \emptyset.$$

The interior and exterior initial-value problems (DM$^\pm$) with mixed boundary conditions consist in finding $u \in C^2(G^\pm) \cap C^1(\bar{G}^\pm)$ satisfying

$$B(\partial_t^2 u)(X) + (Au)(X) = 0, \quad X \in G^\pm,$$
$$u(x, 0+) = (\partial_t u)(x, 0+) = 0, \quad x \in S^\pm,$$
$$u^\pm(X) = f(X), \quad X \in \partial S_1 \times (0, \infty),$$
$$(Tu)^\pm(X) = g(X), \quad X \in \partial S_2 \times (0, \infty).$$

In the interior and exterior initial-boundary value problems $(DC_1^{\pm})$ with combined boundary conditions of the first kind, we look for $u \in C^2(G^{\pm}) \cap C^1(\bar{G}^{\pm})$ such that

$$B(\partial_t^2 u)(X) + (Au)(X) = 0, \quad X \in G^{\pm},$$

$$u(x, 0+) = (\partial_t u)(x, 0+) = 0, \quad x \in S^{\pm},$$

$$u_3^{\pm}(X) = f_3(X), \quad X \in \Gamma,$$

$$(Tu)_{\alpha}^{\pm}(X) = g_{\alpha}(X), \quad X \in \Gamma.$$

If the boundary conditions are of the second kind, then the solution $u \in C^2(G^{\pm}) \cap C^1(\bar{G}^{\pm})$ satisfies

$$B(\partial_t^2 u)(X) + (Au)(X) = 0, \quad X \in G^{\pm},$$

$$u(x, 0+) = (\partial_t u)(x, 0+) = 0, \quad x \in S^{\pm},$$

$$u_{\alpha}^{\pm}(X) = f_{\alpha}(X), \quad X \in \Gamma,$$

$$(Tu)_3^{\pm}(X) = g_3(X), \quad X \in \Gamma.$$

If the regions $S^{\pm} \times [-h_0/2, h_0/2]$ are occupied by two different elastic materials with Lamé constants $\lambda_{\pm}$, $\mu_{\pm}$ and densities $\rho_{\pm}$, respectively, then the initial-boundary value problem (DT) with transmission (contact) boundary conditions consists in finding a pair of functions $u_{\pm} \in C^2(G^{\pm}) \cap C^1(\bar{G}^{\pm})$ such that

$$B_{\pm}(\partial_t^2 u_{\pm})(X) + (A_{\pm} u_{\pm})(X) = 0, \quad X \in G^{\pm},$$

$$u_{\pm}(x, 0+) = (\partial_t u_{\pm})(x, 0+) = 0, \quad x \in S^{\pm},$$

$$u_+^+(X) - u_-^-(X) = f(X), \quad X \in \Gamma,$$

$$(T_+ u_+)^+(X) - (T_- u_-)^-(X) = g(X), \quad X \in \Gamma,$$

where $A_{\pm}$, $B_{\pm}$, and $T_{\pm}$ have the obvious meaning.

Consider an open arc ∂S_0 of ∂S that models a crack, and let

$$\Omega = \mathbb{R}^2 \setminus \overline{\partial S_0},$$

$$\partial S_1 = \partial S \setminus \overline{\partial S_0},$$

$$G = \Omega \times (0, \infty),$$

$$\Gamma_i = \partial S_i \times (0, \infty), \quad i = 0, 1.$$

We write $u \in C^k(\bar{G})$, $k = 0, 1, 2, \ldots$, if the restrictions $u_{\pm}$ of u to $G^{\pm}$ are, respectively, of class $C^k(\bar{G}^{\pm})$ and the limiting values on Γ_1 of u_+ and all its derivatives up to the order k coincide with those of u_-. (These values may differ on Γ_0.) In the initial-boundary value problem (DKD) with Dirichlet boundary conditions, we seek $u \in C^2(G) \cap C^1(\bar{G})$ satisfying

$$B(\partial_t^2 u)(X) + (Au)(X) = 0, \quad X \in G,$$
$$u(x, 0+) = (\partial_t u)(x, 0+) = 0, \quad x \in \Omega,$$
$$u_+^+(X) = f^+(X), \quad X \in \Gamma_0,$$
$$u_-^-(X) = f^-(X), \quad X \in \Gamma_0.$$

The problem (DKN) with Neumann boundary conditions consists in finding $u \in C^2(G) \cap C^1(\bar{G})$ such that

$$B(\partial_t^2 u)(X) + (Au)(X) = 0, \quad X \in G,$$
$$u(x, 0+) = (\partial_t u)(x, 0+) = 0, \quad x \in \Omega,$$
$$(Tu_+)^+(X) = g^+(X), \quad X \in \Gamma_0,$$
$$(Tu_-)^-(X) = g^-(X), \quad X \in \Gamma_0.$$

Let $\mathcal{K}$ be a (3×3)-matrix of the form

$$\mathcal{K} = \begin{pmatrix} \bar{K} & 0 \\ 0 & k_{33} \end{pmatrix},$$

where $k_{33} > 0$ and the (2×2)-matrix $\bar{K} = h^2(k_{\alpha\beta})$ is positive definite. In the interior and exterior initial-boundary value problems $(DD_{\mathcal{K}}^{\pm})$ for a plate on an elastic foundation with Dirichlet boundary conditions, we look for $u \in C^2(G^{\pm}) \cap C^1(\bar{G}^{\pm})$ such that

$$B(\partial_t^2 u)(X) + (Au)(X) + \mathcal{K}u(X) = 0, \quad X \in G^{\pm},$$
$$u(x, 0+) = (\partial_t u)(x, 0+) = 0, \quad x \in S^{\pm},$$
$$u^{\pm}(X) = f(X), \quad X \in \Gamma.$$

The corresponding problems $(DN_{\mathcal{K}}^{\pm})$ with Neumann boundary conditions consist in finding functions $u \in C^2(G^{\pm}) \cap C^1(\bar{G}^{\pm})$ satisfying

$$B(\partial_t^2 u)(X) + (Au)(X) + \mathcal{K}u(X) = 0, \quad X \in G^{\pm},$$
$$u(x, 0+) = (\partial_t u)(x, 0+) = 0, \quad x \in S^{\pm},$$
$$(Tu)^{\pm}(X) = g(X), \quad X \in \Gamma.$$

Throughout what follows, we work frequently with the Laplace transforms of vector-valued functions $u(X) = u(x, t)$, $t \in \mathbb{R}$, which vanish for $t < 0$; that is,

$$\hat{u}(x, p) = \mathcal{L}u(x, t) = \int_0^{\infty} e^{-pt} u(x, t)\, dt.$$

This equality is understood either in the classical or in the distributional sense (see [2], [16], and the Appendix), as the case may be. To simplify the notation, functions or distributions in spaces of transforms will not carry a superposed hat in their symbols when they occur in the general analysis of such spaces; the hat will be added only when there is explicit mention that they are the Laplace transforms of solutions or data occurring in time-dependent problems, or densities of transformed nonstationary (retarded) plate potentials.

We give a brief indication of the technique used to solve the initial-boundary value problems listed above. Roughly speaking, we adopt the following procedure.

(i) Applying the Laplace transformation in (1.9) and taking the homogeneous initial conditions into account, we obtain the equation

$$Bp^2 \hat{u}(x, p) + (A\hat{u})(x, p) = 0, \quad x \in S^+ \text{ or } x \in S^-.$$

Doing the same to the various boundary conditions, we arrive at boundary value problems whose solutions depend on p.

(ii) We prove the unique solvability of the boundary value problems constructed in (i) for any value of p in the complex half-plane

$$\mathbb{C}_\kappa = \{p = \sigma + i\tau \in \mathbb{C} : \sigma > \kappa\}, \quad \kappa > 0,$$

and derive estimates that show how the solutions depend on the complex parameter p.

(iii) Using Parseval's equality, we then return to the spaces of originals and prove the existence of weak solutions to the given initial-boundary value problems.

A similar scheme will also be used to study the solvability of the associated time-dependent boundary integral equations.

1.2 A Matrix of Fundamental Solutions

To construct single-layer and double-layer potentials in the dynamic case, we need a matrix of fundamental solutions $D(X) = D(x, t)$ for (1.9), which vanishes for $t < 0$. In fact, it turns out that we only need the Laplace transform $\hat{D}(x, p)$ of such a matrix because, as shown in [15], for this type of problem numerical methods can be designed that are based solely on estimates for $\hat{D}$ and do not require explicit knowledge of D itself.

Consequently, we seek a (3×3)-matrix $\hat{D}(x, p)$ defined for $p \in \mathbb{C}_0$, which has polynomial growth as $p \to \infty$ and satisfies

$$Bp^2 \hat{D}(x, p) + (A\hat{D})(x, p) = \delta(x)I, \qquad (1.10)$$

where δ is the Dirac distribution and I is the identity (3×3)-matrix. To find $\hat{D}(x, p)$, we apply the (generalized) Fourier transformation with respect to $x \in \mathbb{R}^2$ in (1.10) and arrive at

$$Bp^2 \tilde{D}(\xi, p) + \left(A(\xi)\tilde{D}\right)(\xi, p) = I,$$

or

$$\Theta(\xi, p)\tilde{D}(\xi, p) = I, \tag{1.11}$$

where $\tilde{D}(\xi, p)$ is the Fourier transform of $\hat{D}(x, p)$ and $\Theta(\xi, p)$ is the (3×3)-matrix of elements

$$\Theta_{\alpha\beta}(\xi, p) = h^2(\lambda + \mu)\xi_\alpha\xi_\beta + \delta_{\alpha\beta}(\rho h^2 p^2 + \mu + h^2\mu|\xi|^2),$$
$$\Theta_{33}(\xi, p) = \rho p^2 + \mu|\xi|^2, \tag{1.12}$$
$$\Theta_{\alpha 3}(\xi, p) = -\Theta_{3\alpha}(\xi, p) = i\mu\xi_\alpha.$$

A straightforward calculation yields

$\det \Theta(\xi, p)$
$$= h^4\mu^2(\lambda + 2\mu)|\xi|^6 + h^2\mu\left[\rho p^2 h^2(2\lambda + 5\mu) + \mu(\lambda + 2\mu)\right]|\xi|^4$$
$$+ \rho p^2 h^2(\lambda + 4\mu)(\rho p^2 h^2 + \mu)|\xi|^2 + \rho p^2(\rho p^2 h^2 + \mu)^2.$$

From this it follows that $\det \Theta(\xi, p)$ is invariant with respect to rotations in $\mathbb{R}^2$; that is, $\det \Theta(\xi, p)$ depends only on $|\xi|$. Then we take $\xi_1 = |\xi|$ and $\xi_2 = 0$ in (1.12) and find that

$\det \Theta(\xi, p) = R(|\xi|^2, p)$

$$= \begin{vmatrix} \rho p^2 h^2 + \mu + h^2(\lambda + 2\mu)|\xi|^2 & 0 & -i\mu|\xi| \\ 0 & \rho p^2 h^2 + \mu + h^2\mu|\xi|^2 & 0 \\ i\mu|\xi| & 0 & \rho p^2 + \mu|\xi|^2 \end{vmatrix}$$

$$= (h^2\mu|\xi|^2 + \rho p^2 h^2 + \mu)\left[h^2\mu(\lambda + 2\mu)|\xi|^4 \right.$$
$$\left. + \rho p^2 h^2(\lambda + 3\mu)|\xi|^2 + \rho p^2(\rho p^2 h^2 + \mu)\right].$$

Let $|\xi|^2 = s$. We denote the roots of the equation $R(s, p) = 0$ by $s_i = -\chi_i^2$ and choose χ_i so that $\operatorname{Re}\chi_i \geq 0$. It is easy to check that

$$\chi_{1,2}^2 = p\left[2h\mu(\lambda + 2\mu)\right]^{-1}$$
$$\times \left\{\rho p h(\lambda + 3\mu) \pm \left[\rho^2 p^2 h^2(\lambda + \mu)^2 - 4\rho\mu^2(\lambda + 2\mu)\right]^{1/2}\right\},$$
$$\chi_3^2 = (h^2\mu)^{-1}(\rho p^2 h^2 + \mu).$$

1.1 Lemma. (i) *The equation $R(s, p) = 0$ does not have a triple root for any $p \in \mathbb{C}_0$.*

(ii) *The equation $R(s, p) = 0$ has a double root if and only if*

$$p = 2\mu \big[h(\lambda + \mu) \big]^{-1} \big[\rho^{-1}(\lambda + 2\mu) \big]^{1/2}.$$

In this case, $\chi_1^2 = \chi_2^2$.

(iii) *$\operatorname{Re} \chi_i > 0$ for every $p \in \mathbb{C}_0$.*

Proof. (i) Suppose that $\chi_1^2 = \chi_2^2 = \chi_3^2$. Then the explicit expressions of the roots show that $\chi_1^2 = \chi_2^2$ implies that

$$\rho^2 p^2 h^2 (\lambda + \mu)^2 - 4\rho\mu^2(\lambda + 2\mu) = 0;$$

hence,

$$p = 2\mu \big[h(\lambda + \mu) \big]^{-1} \big[\rho^{-1}(\lambda + 2\mu) \big]^{1/2}.$$

Now the equality $\chi_1^2 = \chi_3^2$ yields $\lambda + \mu = 0$, which contradicts (1.1).

(ii) If $\chi_1^2 = \chi_3^2$ or $\chi_2^2 = \chi_3^2$, then, as a straightforward calculation shows, we obtain

$$p^2 = -\mu(\rho h^2)^{-1},$$

which is impossible, since $\operatorname{Re} p > 0$.

(iii) First, we note that $\chi_i \neq 0$. If $\operatorname{Re} \chi_i = 0$ for some i, then $\chi_i^2 < 0$; hence, $\tilde{s} = |\tilde{\xi}|^2 = -\chi_i^2$ is a positive root of the equation $R(s, p) = 0$. We take $\tilde{\xi} = (|\tilde{\xi}|, 0)$ and denote by $g = (g_1, g_2, g_3)^{\mathrm{T}}$ a nonzero solution of the system of linear algebraic equations $\Theta(\tilde{\xi}, p)g = 0$. Multiplying this equality by g^*, we obtain

$$(\rho h^2 p^2 + \mu)(|g_1|^2 + |g_2|^2) + (\rho p^2 + \mu|\tilde{\xi}|^2)|g_3|^2$$
$$+ h^2 |\tilde{\xi}|^2 \big[(\lambda + 2\mu)|g_1|^2 + \mu|g_2|^2 \big] - 2\mu|\tilde{\xi}| \operatorname{Re}(g_3^* i g_1) = 0;$$

consequently, $p^2 \in \mathbb{R}$ and $p^2 > 0$. Since

$$2\mu|\tilde{\xi}| \operatorname{Re}(g_3^* i g_1) \geq -\mu(|g_1|^2 + |\tilde{\xi}|^2|g_3|^2),$$

it follows that

$$\rho p^2 h^2 (|g_1|^2 + |g_2|^2) + \rho p^2 |g_3|^2$$
$$+ h^2 |\tilde{\xi}|^2 \big[(\lambda + 2\mu)|g_1|^2 + \mu|g_2|^2 \big] \leq 0.$$

This contradiction completes the proof.   □

We write

$$\tilde{\Psi}(\xi, p) = \big[\det \Theta(\xi, p) \big]^{-1}.$$

By (1.11) and (1.12),

$$\tilde{D}_{11}(\xi, p) = \big[h^2\mu^2|\xi|^4 + h^2\mu(\lambda + \mu)|\xi|^2\xi_2^2 + 2\rho p^2 h^2\mu|\xi|^2 + \mu^2\xi_1^2$$
$$+ \rho p^2 h^2(\lambda + \mu)\xi_2^2 + \rho p^2(\rho p^2 h^2 + \mu)\big]\tilde{\Psi}(\xi, p),$$
$$\tilde{D}_{22}(\xi, p) = \big[h^2\mu^2|\xi|^4 + h^2\mu(\lambda + \mu)|\xi|^2\xi_1^2 + 2\rho p^2 h^2\mu|\xi|^2 + \mu^2\xi_2^2$$
$$+ \rho p^2 h^2(\lambda + \mu)\xi_1^2 + \rho p^2(\rho p^2 h^2 + \mu)\big]\tilde{\Psi}(\xi, p),$$
$$\tilde{D}_{33}(\xi, p) = \big[h^4\mu(\lambda + 2\mu)|\xi|^4 + h^2(\rho p^2 h^2 + \mu)(\lambda + 3\mu)|\xi|^2$$
$$+ (\rho p^2 h^2 + \mu)^2\big]\tilde{\Psi}(\xi, p),$$
$$\tilde{D}_{12}(\xi, p) = \tilde{D}_{21}(\xi, p) = -\xi_1\xi_2\big[h^2\mu(\lambda + \mu)|\xi|^2$$
$$+ \rho p^2 h^2(\lambda + \mu) - \mu^2\big]\tilde{\Psi}(\xi, p),$$
$$\tilde{D}_{\alpha 3}(\xi, p) = -\tilde{D}_{3\alpha}(\xi, p) = i\mu\xi_\alpha(h^2\mu|\xi|^2 + \rho p^2 h^2 + \mu)\tilde{\Psi}(\xi, p).$$

Hence, the elements of the matrix of fundamental solutions $\hat{D}(x, p)$ are

$$\hat{D}_{11}(x, p) = \big[h^2\mu^2\Delta^2 + h^2\mu(\lambda + \mu)\Delta\partial_2^2 - 2\rho p^2 h^2\mu\Delta - \mu^2\partial_1^2$$
$$- \rho p^2 h^2(\lambda + \mu)\partial_2^2 + \rho p^2(\rho p^2 h^2 + \mu)\big]\Psi(x, p),$$
$$\hat{D}_{22}(x, p) = \big[h^2\mu^2\Delta^2 + h^2\mu(\lambda + \mu)\Delta\partial_1^2 - 2\rho p^2 h^2\mu\Delta - \mu^2\partial_2^2$$
$$- \rho p^2 h^2(\lambda + \mu)\partial_1^2 + \rho p^2(\rho p^2 h^2 + \mu)\big]\Psi(x, p),$$
$$\hat{D}_{33}(x, p) = \big[h^4\mu(\lambda + 2\mu)\Delta^2 - h^2(\rho p^2 h^2 + \mu)(\lambda + 3\mu)\Delta$$
$$+ (\rho p^2 h^2 + \mu)^2\big]\Psi(x, p), \tag{1.13}$$
$$\hat{D}_{12}(x, p) = \hat{D}_{21}(x, p) = \big[-h^2\mu(\lambda + \mu)\Delta$$
$$+ \rho p^2 h^2(\lambda + \mu) - \mu^2\big]\partial_1\partial_2\Psi(x, p),$$
$$\hat{D}_{\alpha 3}(x, p) = -\hat{D}_{3\alpha}(x, p) = \mu(h^2\mu\Delta - \rho p^2 h^2 - \mu)\partial_\alpha\Psi(x, p),$$

where $\Psi(x, p)$ is the inverse Fourier transform of $\tilde{\Psi}(\xi, p)$; that is,

$$\Psi(x, p) = (4\pi^2)^{-1}\int_{\mathbb{R}^2} e^{-i(x,\xi)}\tilde{\Psi}(\xi, p)\, d\xi.$$

By Lemma 1.1, there are two possible cases. In the first one, the roots of the equation $R(s, p) = 0$ are simple. Direct calculation shows that in this case

$$\Psi(x, p) = \big[2\pi h^4\mu^2(\lambda + 2\mu)\big]^{-1} c_i K_0(\chi_i|x|), \tag{1.14}$$

where K_0 is the modified Bessel function of order zero [1] and

$$c_1 = [(\chi_1^2 - \chi_2^2)(\chi_1^2 - \chi_3^2)]^{-1},$$
$$c_2 = [(\chi_2^2 - \chi_3^2)(\chi_2^2 - \chi_1^2)]^{-1}, \qquad (1.15)$$
$$c_3 = [(\chi_3^2 - \chi_1^2)(\chi_3^2 - \chi_2^2)]^{-1}.$$

In the second case, the equation $R(s,p) = 0$ has a double root:

$$\chi_1^2 = \chi_2^2 \neq \chi_3^2.$$

Then

$$\Psi(x,p) = \left[2\pi h^4 \mu^2(\lambda + 2\mu)\right]^{-1}\left[\tilde{c}_1 K_0(\chi_1|x|)\right.$$
$$\left. + \tilde{c}_2 (2\chi_1)^{-1}|x|K_1(\chi_1|x|) + \tilde{c}_3 K_0(\chi_3|x|)\right], \qquad (1.16)$$

where K_1 is the modified Bessel function of order one and

$$\tilde{c}_1 = -\tilde{c}_3 = -(\chi_3^2 - \chi_1^2)^{-2},$$
$$\tilde{c}_2 = (\chi_3^2 - \chi_1^2)^{-1}. \qquad (1.17)$$

1.2 Lemma. *For any $p \in \mathbb{C}_0$, the function $\Psi(x,p)$ can be represented in the neighborhood of $x = 0$ in the form*

$$\Psi(x,p) = -\left[128\pi h^4 \mu^2(\lambda + 2\mu)\right]^{-1}|x|^4 \ln|x|$$
$$+ O(|x|^6 \ln|x|) + \Psi_0(x,p), \qquad (1.18)$$

where $\Psi_0(x,p)$ is an infinitely differentiable function; in addition, $\Psi(x,p) \to 0$ exponentially as $|x| \to \infty$.

Proof. In the case of simple roots, from (1.14) and the asymptotic behavior of the modified Bessel function $K_0(z)$ as $z \to 0$ [1] it follows that

$$\Psi(x,p) = -\left[2\pi h^4 \mu^2(\lambda + 2\mu)\right]^{-1} \ln|x|$$
$$+ \left[c_1 + c_2 + c_3 + c_i\left(\tfrac{1}{4}\chi_i^2|x|^2 + \tfrac{1}{64}\chi_i^4|x|^4\right)\right]\ln|x|$$
$$+ O(|x|^6 \ln|x|) + \Psi_0(x,p).$$

Using (1.15), it is easy to verify that

$$c_1 + c_2 + c_3 = 0,$$
$$c_i\chi_i^2 = 0,$$
$$c_i\chi_i^4 = 1;$$

therefore, (1.18) holds.

In the case of a double root, (1.16) implies that

$$\Psi(x,p) = -\left[2\pi h^4 \mu^2 (\lambda + 2\mu)\right]^{-1} \ln|x|$$
$$\times \left[\tilde{c}_1 \left(1 + \tfrac{1}{4}\chi_1^2 |x|^2 + \tfrac{1}{64}\chi_1^4 |x|^4\right) - \tfrac{1}{4}\tilde{c}_1 |x|^2 \left(1 + \tfrac{1}{8}\chi_1^2 |x|^2\right)\right.$$
$$\left. + \tilde{c}_3 \left(1 + \tfrac{1}{4}\chi_3^2 |x|^2 + \tfrac{1}{64}\chi_3^4 |x|^4\right)\right] \ln|x|$$
$$+ O(|x|^6 \ln|x|) + \Psi_0(x,p),$$

and we immediately regain (1.18) from (1.17).

The last assertion follows from the fact that $\mathrm{Re}\,\chi_i > 0$ and the asymptotic behavior of $K_n(z)$, $n = 0, 1$, as $z \to \infty$ in such a way that $\mathrm{Re}\,z \geq \kappa > 0$.   □

1.3 Corollary. *For any $p \in \mathbb{C}_0$, the elements of the matrix of fundamental solutions $\hat{D}(x,p)$ can be represented in the neighborhood of $x = 0$ in the form*

$$\hat{D}_{\alpha\beta}(x,p) = \left[4\pi\mu(\lambda + 2\mu)h^2\right]^{-1}$$
$$\times \left[(\lambda + \mu)x_\alpha x_\beta |x|^{-2} - (\lambda + 3\mu)\delta_{\alpha\beta} \ln|x|\right]$$
$$+ O(|x|^2 \ln|x|) + \hat{D}_{0,\alpha\beta}(x,p),$$
$$\hat{D}_{33}(x,p) = -(2\pi h^2 \mu)^{-1} \ln|x| \tag{1.19}$$
$$+ O(|x|^2 \ln|x|) + \hat{D}_{0,33}(x,p),$$
$$\hat{D}_{\alpha3}(x,p) = -\hat{D}_{3\alpha}(x,p) = -\left[4\pi h^2 (\lambda + 2\mu)\right]^{-1} x_\alpha \ln|x|$$
$$+ O(|x|^2 \ln|x|) + \hat{D}_{0,\alpha3}(x,p),$$

where $\hat{D}_{0,ij}(x,p)$ are infinitely differentiable functions.

The proof of this assertion follows from (1.13) and (1.18).

1.4 Remark. Representation (1.19) shows that for any $p \in \mathbb{C}_0$, the asymptotic behavior of $\hat{D}(x,p)$ in the neighborhood of $x = 0$ coincides with that of the matrix of fundamental solutions $D(x)$ for the equilibrium equation [10].

1.3 Time-dependent Plate Potentials

Let α, $\beta \in C^2(\partial S \times \mathbb{R})$ be three-component vector-valued functions with compact support in $\bar{\Gamma}$. First, we introduce single-layer and double-layer potentials in terms of Laplace transforms. Thus, if

$$\hat{\alpha}(x,p) = \mathcal{L}\alpha(x,t), \quad \hat{\beta}(x,p) = \mathcal{L}\beta(x,t),$$

then for every $p \in \mathbb{C}_0$ we define a single-layer potential $V_p\hat{\alpha}$ of density $\hat{\alpha}$ by

$$(V_p\hat{\alpha})(x,p) = \int_{\partial S} \hat{D}(x - y, p)\hat{\alpha}(y, p)\, ds_y, \quad x \in \mathbb{R}^2,$$

and a double-layer potential $W_p\hat{\beta}$ of density $\hat{\beta}$ by

$$(W_p\hat{\beta})(x,p) = \int_{\partial S} \hat{P}(x - y, p)\hat{\beta}(y, p)\, ds_y$$

$$= \int_{\partial S} (\hat{\beta}(y, p), T_y\hat{D}^{(j)}(y - x, p))e_j\, ds_y, \quad x \in S^+ \cup S^-,$$

where

$$\hat{P}(x - y, p) = \left[T_y\hat{D}(y - x, p)\right]^{\mathrm{T}},$$

T_y is the moment-force boundary operator acting with respect to the point y, and e_j is the jth coordinate unit vector in $\mathbb{R}^3$.

By Remark 1.4, for any fixed $p \in \mathbb{C}_0$ the boundary properties of both potentials coincide with those of the corresponding single-layer and double-layer potentials in the equilibrium case [10]. Below we list the most significant of these properties.

(i) The single-layer potential $V_p\hat{\alpha}$ satisfies the equation

$$Bp^2\hat{u}(x, p) + (A\hat{u})(x, p) = 0, \quad x \in S^+ \cup S^-. \tag{1.20}$$

(ii) $V_p\hat{\alpha}$ is continuous in $\mathbb{R}^2$; in particular, the direct value $(V_p\hat{\alpha})_0$ on ∂S of the corresponding weakly singular integral is given by

$$(V_p\hat{\alpha})_0(x, p) = (V_p\hat{\alpha})^+(x, p) = (V_p\hat{\alpha})^-(x, p), \quad x \in \partial S.$$

(iii) There hold the jump formulas

$$(TV_p\hat{\alpha})^\pm(x, p) = \pm\tfrac{1}{2}\hat{\alpha}(x, p) + (TV_p\hat{\alpha})_0(x, p), \quad x \in \partial S,$$

where $(TV_p\hat{\alpha})_0$ is the direct value of the corresponding singular integral on ∂S.

(iv) The double-layer potential $W_p\hat{\beta}$ satisfies (1.20) and can be extended by continuity from $S^\pm$ to $\bar{S}^\pm$, respectively. These extensions are of class $C^{1,\beta}(\bar{S}^\pm)$ for any $\beta \in (0, 1)$.

(v) There hold the jump formulas

$$(W_p\hat{\beta})^\pm(x, p) = \mp\tfrac{1}{2}\hat{\beta}(x, p) + (W_p\hat{\beta})_0(x, p), \quad x \in \partial S,$$

where $(W_p\hat{\beta})_0$ is the direct value of the corresponding singular integral on ∂S.

(vi) There holds the equality

$$(TW_p\hat{\beta})^+(x,p) = (TW_p\hat{\beta})^-(x,p), \quad x \in \partial S.$$

This equality enables us to introduce the notation

$$(N_p\hat{\beta})(x,p) = (TW_p\hat{\beta})^+(x,p) = (TW_p\hat{\beta})^-(x,p), \quad x \in \partial S.$$

Since $\hat{D}(x,p)$ has polynomial growth with respect to $p \in \mathbb{C}_\kappa$, $\kappa > 0$, we can define the single-layer and double-layer time-dependent (retarded) potentials $V\alpha$ and $W\beta$ as the inverse Laplace transforms of $V_p\hat{\alpha}$ and $W_p\hat{\beta}$, respectively; more precisely, given that the inverse transform of a product of two transforms is the convolution of the corresponding two originals, we have

$$(V\alpha)(X) = (\mathcal{L}^{-1}V_p\hat{\alpha})(X)$$
$$= \int_0^\infty \int_{\partial S} D(x-y,t-\tau)\alpha(y,\tau)\,ds_y\,d\tau,$$

$$(W\beta)(X) = (\mathcal{L}^{-1}W_p\hat{\beta})(X)$$
$$= \int_0^\infty \int_{\partial S} P(x-y,t-\tau)\beta(y,\tau)\,ds_y\,d\tau,$$

where $P(x-y,t)$ is the inverse Laplace transform of $\hat{P}(x-y,p)$; that is,

$$P(x-y,t-\tau)\beta(y,\tau) = \big(\beta(y,\tau), T_y D^{(j)}(y-x,t-\tau)\big)e_j.$$

From the properties of $V_p\hat{\alpha}$ and $W_p\hat{\beta}$ it follows that, at least for smooth densities α and β vanishing for $t < 0$ and with compact support with respect to the time variable, the retarded potentials have the following properties.

(i) $V\alpha$ and $W\beta$ satisfy the equation

$$B(\partial_t^2 u)(X) + (Au)(X) = 0, \quad X \in G^+ \cup G^-.$$

(ii) $V\alpha$ and $W\beta$ satisfy the initial conditions

$$u(x,0+) = (\partial_t u)(x,0+) = 0, \quad x \in S^+ \cup S^-.$$

(iii) For $X \in \Gamma$, there hold the jump formulas

$$(V\alpha)^+(X) = (V\alpha)^-(X),$$
$$(TV\alpha)^+(X) - (TV\alpha)^-(X) = \alpha(X),$$
$$(W\beta)^+(X) - (W\beta)^-(X) = -\beta(X),$$
$$(TW\beta)^+(X) = (TW\beta)^-(X).$$

After we introduce appropriate Sobolev-type spaces and study the properties of the boundary operators generated by the potentials in these spaces, we show that the above jump formulas hold, in fact, for much wider classes of densities α and β if the limiting values of the potentials are understood as traces on Γ.

1.4 Nonstationary Boundary Integral Equations

We intend to seek integral representations for the solutions of all the initial-boundary value problems listed in §1.1. As a typical example, here we discuss such representations for $(DD^{\pm})$ and $(DN^{\pm})$.

Suppose that we want to represent the solutions of $(DD^{\pm})$ in the form

$$u(X) = (V\alpha)(X), \quad X \in G^{\pm}, \tag{1.21}$$

where α is an unknown density. Since, as indicated in §1.3, $V\alpha$ satisfies both the equation of motion and the initial conditions, u needs to satisfy only the Dirichlet boundary conditions, which means that α must be a solution of the nonstationary boundary integral equation

$$(V\alpha)(X) = f(X), \quad X \in \Gamma. \tag{1.22}$$

The kernel $D(X - Y)$ of this equation has a retarded time argument. For every fixed value of this argument, the kernel is weakly singular with respect to the space variables.

If we now seek the solutions of $(DD^{\pm})$ in the form

$$u(X) = (W\beta)(X), \quad X \in G^{\pm}, \tag{1.23}$$

then, according to the properties of $W\beta$, the unknown density β must satisfy the corresponding boundary integral equation

$$(W\beta)^{\pm}(X) = f(X), \quad X \in \Gamma, \tag{1.24}$$

whose kernel again has a retarded time argument but is singular with respect to the space variables.

In the initial-boundary value problems $(DN^{\pm})$, the representations (1.21) and (1.23) yield, respectively, the nonstationary boundary integral equations

$$(TV\alpha)^{\pm}(X) = g(X), \quad X \in \Gamma, \tag{1.25}$$

$$(N\beta)(X) = g(X), \quad X \in \Gamma, \tag{1.26}$$

where $N\beta$ is the inverse Laplace transform of $N_p\hat{\beta}$. The kernel of (1.25) has the same structure as that of (1.24), while the kernel of (1.26) is hypersingular with respect to the space variables. Both kernels have a retarded time argument.

Comparing the representations (1.21) and (1.23) with the corresponding ones in the equilibrium problems [10], we see that here we have no additive rigid displacements. This remark is valid for the integral representations of the solutions to all the initial-boundary value problems that we are considering.

In what follows, we apply the Laplace transformation with respect to time in the nonstationary boundary integral equations and reduce them to new boundary integral equations that depend on the transformation parameter p. The main difficulty in the study of the latter resides in establishing estimates for their solutions that illustrate the nature of the dependence of these solutions on p. Once such estimates are obtained, we can prove the unique solvability of the time-dependent boundary equations.

2

Problems with Dirichlet Boundary Conditions

2.1 Function Spaces

In what follows, we use the same notation for norms and inner products on spaces of scalar and vector-valued functions, in accordance with the convention adopted in §1.1. We also point out that all these function spaces are complex.

We start by introducing spaces of functions that depend on a complex parameter p; their properties, studied in [2], are listed below without proof.

Let $m \in \mathbb{R}$, $p \in \mathbb{C}$, $k \in \mathbb{R}$, and $S \subset \mathbb{R}^2$.

$H_m(\mathbb{R}^2)$, $H_m(S)$, and $H_m(\partial S)$ are the standard Sobolev spaces whose elements are defined on $\mathbb{R}^2$, S, and the boundary ∂S of S, respectively.

$H_{m,p}(\mathbb{R}^2)$ is the space that coincides as a set with $H_m(\mathbb{R}^2)$ but is equipped with the norm

$$\|u\|_{m,p} = \left\{ \int_{\mathbb{R}^2} (1 + |p|^2 + |\xi|^2)^m |\tilde{u}(\xi)|^2 \, d\xi \right\}^{1/2},$$

where $\tilde{u}$ is the (distributional) Fourier transform of the three-component distribution $u \in \mathcal{S}'(\mathbb{R}^2)$ (see the Appendix). Clearly, for any fixed $p \in \mathbb{C}$, the norms on $H_{m,p}(\mathbb{R}^2)$ and $H_m(\mathbb{R}^2)$ are equivalent.

$\mathring{H}_{m,p}(S)$ is the subspace of all $u \in H_{m,p}(\mathbb{R}^2)$ such that $\operatorname{supp} u \subset \bar{S}$.

$H_{m,p}(S)$ is the space of the restrictions to S of all $v \in H_{m,p}(\mathbb{R}^2)$. The norm of $u \in H_{m,p}(S)$ is defined by

$$\|u\|_{m,p;S} = \inf_{v \in H_{m,p}(\mathbb{R}^2):\, v|_S = u} \|v\|_{m,p}.$$

The inner products in $L^2(\mathbb{R}^2)$, $L^2(S)$, and $L^2(\partial S)$ are denoted by $(\cdot,\cdot)_0$, $(\cdot,\cdot)_{0;S}$, and $(\cdot,\cdot)_{0;\partial S}$, respectively.

$H_{-m,p}(\mathbb{R}^2)$ is the dual of $H_{m,p}(\mathbb{R}^2)$ with respect to the duality generated by $(\cdot,\cdot)_0$.

$H_{-m,p}(S)$ is the dual of $\mathring{H}_{m,p}(S)$.

$H_{1/2,p}(\partial S)$ is the space of the traces on ∂S of all the elements of $H_{1,p}(S)$. It coincides as a set with $H_{1/2}(\partial S)$ but is equipped with the norm

$$\|f\|_{1/2,p;\partial S} = \inf_{u \in H_{1,p}(S): \gamma u = f} \|u\|_{1,p;S}.$$

Here γ is the trace operator, which maps $H_{1,p}(S)$ continuously to $H_{1/2,p}(\partial S)$. We mention that the continuity of γ is uniform with respect to $p \in \mathbb{C}$; that is,

$$\|\gamma u\|_{1/2,p;\partial S} \leq c\|u\|_{1,p;S},$$

where c does not depend on $p \in \mathbb{C}$ [2]. We recall that the trace operators corresponding to the interior and exterior domains $S^{\pm}$ are denoted by $\gamma^{\pm}$.

$H_{-1/2,p}(\partial S)$ is the dual of $H_{1/2,p}(\partial S)$ with respect to the duality generated by $(\cdot\,,\cdot)_{0;\partial S}$.

Next, l^+ and l^- are extension operators that, in the context of our function spaces, map $H_{1/2,p}(\partial S)$ to $H_{1,p}(S^+)$ and $H_{1,p}(S^-)$ continuously and uniformly with respect to $p \in \mathbb{C}$.

$H_{m,k,\kappa}(S)$ is the space of all $\hat{u}(x,p)$, $x \in S$, $p \in \mathbb{C}_{\kappa}$, such that the mapping $U(p) = \hat{u}(\cdot\,,p)$ is holomorphic from $\mathbb{C}_{\kappa}$ to $H_m(S)$ and

$$\|\hat{u}\|^2_{m,k,\kappa;S} = \sup_{\sigma > \kappa} \int_{-\infty}^{\infty} (1 + |p|^2)^k \|U(p)\|^2_{m,p;S}\, d\tau < \infty, \quad p = \sigma + i\tau. \quad (2.1)$$

Formula (2.1) defines the norm on $H_{m,k,\kappa}(S)$. It is readily seen from its definition that $U(p) \in H_{m,p}(S)$ for any $p \in \mathbb{C}_{\kappa}$. In what follows, we write $\hat{u}(x,p)$ if we want to emphasize that this is an element of $H_{m,p}(S)$, and $U(p)$ when we want to regard it as a mapping from $\mathbb{C}_{\kappa}$ to $H_m(S)$.

$H_{\pm 1/2,k,\kappa}(\partial S)$ are introduced similarly; that is, these spaces consist of all $\hat{f}(x,p)$, $x \in \partial S$, $p \in \mathbb{C}_{\kappa}$, such that the corresponding mapping $F(p) = \hat{f}(\cdot\,,p)$ is holomorphic from $\mathbb{C}_{\kappa}$ to $H_{\pm 1/2}(\partial S)$ and

$$\|\hat{f}\|^2_{\pm 1/2,k,\kappa;\partial S} = \sup_{\sigma > \kappa} \int_{-\infty}^{\infty} (1 + |p|^2)^k \|F(p)\|^2_{\pm 1/2,p;\partial S}\, d\tau < \infty.$$

Once again, the above equality defines the norms on these spaces and, as above, we interpret $\hat{f}(x,p)$ as an element of $H_{\pm 1/2,p}(\partial S)$ and $F(p)$ as a mapping from $\mathbb{C}_{\kappa}$ to $H_{\pm 1/2}(\partial S)$.

$H_{m,k,\kappa}(G)$ and $H_{\pm 1/2,k,\kappa}(\Gamma)$ consist, respectively, of the inverse Laplace transforms u and f of the elements $\hat{u}$ and $\hat{f}$ of $H_{m,k,\kappa}(S)$ and $H_{\pm 1/2,k,\kappa}(\partial S)$; these spaces are equipped with the norms

$$\|u\|_{m,k,\kappa;G} = \|\hat{u}\|_{m,k,\kappa;S},$$

$$\|f\|_{\pm 1/2,k,\kappa;\Gamma} = \|\hat{f}\|_{\pm 1/2,k,\kappa;\partial S}.$$

$$(2.2)$$

By the Paley–Wiener theorem and Parseval's equality [12], for a nonnegative integer k the spaces $H_{1,k,\kappa}(G)$ consist of all three-component distributions u defined on $S \times \mathbb{R}$ that vanish for $t < 0$ and are such that

$$\int_G e^{-2\kappa t} \sum_{|\alpha|+\alpha_t \leq 1} |(\partial_x^\alpha \partial_t^{\alpha_t+k} u)(x,t)|^2 \, dx \, dt < \infty, \qquad (2.3)$$

where α is a two-component multi-index, α_t is a nonnegative integer, and ∂_x^α is the partial differentiation operator acting with respect to the space variables. The norm on $H_{1,k,\kappa}(G)$ defined by (2.3) is equivalent to (2.2). A similar remark is also valid for $H_{1/2,k,\kappa}(\Gamma)$.

In what follows, we relax the terminology and refer to the elements of all of the above spaces as "functions" instead of "distributions" or "generalized functions", since the former term, although technically incorrect, is more familiar to the nonspecialist reader.

Finally, $C_0^\infty(\bar{G}^\pm)$ are the spaces of infinitely differentiable functions with compact support in $\bar{G}^\pm$, respectively.

2.2 Solvability of the Transformed Problems

In this section, we discuss the problems $(\mathrm{D}_p^\pm)$ obtained after applying the Laplace transformation with respect to the time variable in the original problems $(\mathrm{DD}^\pm)$. Our aim is to establish the unique solvability of $(\mathrm{D}_p^\pm)$ for every $p \in \mathbb{C}_0$ and derive certain estimates for their solutions. We use two different approaches to solve these problems. The first one is based on the Fredholm Alternative and works well in the case of interior domains. Unfortunately, the same cannot be done in exterior problems because here the operators occurring in the corresponding functional equations lose their compactness in the natural spaces where such problems are set. This makes it necessary for us to modify the method when we deal with exterior domains.

The transformed problems that we consider here are more general in that they include the contribution of the body forces and moments q occurring on the right-hand side in the equation of motion (1.8). Thus, the classical version of these more general problems $(\mathrm{D}_p^\pm)$ consists in finding $\hat{u} \in C^2(S^\pm) \cap C(\bar{S}^\pm)$ such that

$$Bp^2 \hat{u}(x,p) + (A\hat{u})(x,p) = \hat{q}(x,p), \quad x \in S^\pm,$$
$$\hat{u}^\pm(x,p) = \hat{f}(x,p), \quad x \in \partial S, \qquad (2.4)$$

where

$$\hat{u}(x,p) = \mathcal{L}u(x,t), \quad \hat{q}(x,p) = \mathcal{L}q(x,t), \quad \hat{f}(x,p) = \mathcal{L}f(x,t).$$

In order to simplify the notation, and since there is no danger of ambiguity, throughout this section we omit the hat from the symbols of functions that depend on x and p, it being understood that, unless otherwise stipulated, we are carrying out our arguments in spaces of Laplace transforms.

As usual, to derive the variational version of $(D_p^{\pm})$, we multiply (2.4) termwise by $v^* \in C_0^{\infty}(S^{\pm})$ and integrate the result over $S^{\pm}$; thus, we obtain

$$p^2(Bu, v)_{0;S^{\pm}} + a_{\pm}(u, v) = (q, v)_{0;S^{\pm}}, \tag{2.5}$$

where

$$(Bu, v)_{0;S^{\pm}} = (B^{1/2}u, B^{1/2}v)_{0;S^{\pm}},$$

$$a_{\pm}(u, v) = 2 \int_{S^{\pm}} E(u, v^{*\mathrm{T}}) \, dx,$$

and $E(u, v)$ is the sesquilinear form defined by the internal energy density; that is [9],

$$2E(u, v) = h^2 E_0(u, v) + h^2 \mu(\partial_2 u_1 + \partial_1 u_2)(\partial_2 \bar{v}_1 + \partial_1 \bar{v}_2)$$

$$+ \mu[(u_1 + \partial_1 u_3)(\bar{v}_1 + \partial_1 \bar{v}_3) + (u_2 + \partial_2 u_3)(\bar{v}_2 + \partial_2 \bar{v}_3)],$$

$$E_0(u, v) = (\lambda + 2\mu)\big[(\partial_1 u_1)(\partial_1 \bar{v}_1) + (\partial_2 u_2)(\partial_2 \bar{v}_2)\big]$$

$$+ \lambda\big[(\partial_1 u_1)(\partial_2 \bar{v}_2) + (\partial_2 u_2)(\partial_1 \bar{v}_1)\big].$$

It is obvious that

$$a_{\pm}(u, v) = \overline{a_{\pm}(v, u)}, \tag{2.6}$$

where the superposed bar denotes complex conjugation. Equation (2.5) indicates that the variational version of problems (2.4) should consist in finding $u \in H_{1,p}(S^{\pm})$ such that

$$p^2(B^{1/2}u, B^{1/2}v)_{0;S^{\pm}} + a_{\pm}(u, v) = (q, v)_{0;S^{\pm}} \quad \forall v \in \mathring{H}_{1,p}(S^{\pm}),$$
$$\gamma^{\pm}u = f. \tag{2.7}$$

Throughout what follows, we denote by the same symbol c all positive constants occurring in estimates, which are independent of the functions in these estimates and of $p \in \mathbb{C}_{\kappa}$, but may depend on κ.

2.1 Theorem. *For any $f \in H_{1/2,p}(\partial S)$ and $q \in H_{-1,p}(S^+)$, $p \in \bar{\mathbb{C}}_{\kappa}$, $\kappa > 0$, problem (D_p^+) has a unique weak solution $u \in H_{1,p}(S^+)$ and*

$$\|u\|_{1,p;S^+} \le c|p|(\|q\|_{-1,p;S^+} + \|f\|_{1/2,p;\partial S}). \tag{2.8}$$

Proof. First we consider (D_p^+) with homogeneous boundary conditions, which consists in finding $u_0 \in \mathring{H}_{1,p}(S^+)$ such that

$$p^2(B^{1/2}u_0, B^{1/2}v)_{0;S^+} + a_+(u_0, v) = (q, v)_{0;S^+} \quad \forall v \in \mathring{H}_{1,p}(S^+). \tag{2.9}$$

Repeating the proof of Lemma 2.3 in [7], we can show that $a_+(u, v)$ is coercive on $[\mathring{H}_{1,p}(S^+)]^2$. Since the form is also continuous on this space, we conclude that for any $q \in H_{-1,p}(S^+)$, the variational equation

$$a_+(u_0, v) = (q, v)_{0;S^+} \quad \forall v \in \mathring{H}_{1,p}(S^+) \tag{2.10}$$

has a unique solution $u_0 \in \mathring{H}_{1,p}(S^+)$, which satisfies the estimate

$$\|u_0\|_1 \le c\|q\|_{-1;S^+}.$$

On the other hand, since $a_+(u_0, v)$ defines a bounded antilinear (conjugate linear) functional on $\mathring{H}_{1,p}(S^+)$ for any $u_0 \in \mathring{H}_{1,p}(S^+)$, it can be written in the form (2.10) with some $q \in H_{-1,p}(S^+)$. Let $\mathcal{A}$ be the operator defined by $a_+(u_0, v)$, which associates $u_0 \in \mathring{H}_{1,p}(S^+)$ with $q \in H_{-1,p}(S^+)$ as described above; thus,

$$a_+(u_0, v) = (\mathcal{A}u_0, v)_{0;S^+} \quad \forall u_0, v \in \mathring{H}_1(S^+).$$

$\mathcal{A}$ is a homeomorphism from $\mathring{H}_1(S^+)$ to $H_{-1}(S^+)$. Additionally, from (2.6) it follows that $\mathcal{A}$ is self-adjoint in the sense that

$$(\mathcal{A}u_0, v)_{0;S^+} = (u_0, \mathcal{A}v)_{0;S^+} \quad \forall u_0, v \in \mathring{H}_{1,p}(S^+).$$

This is easily verified, since for any $u_0, v \in \mathring{H}_{1,p}(S^+)$,

$$(\mathcal{A}u_0, v)_{0;S^+} = a_+(u_0, v) = \overline{a_+(v, u_0)}$$

$$= \overline{(\mathcal{A}v, u_0)_{0;S^+}} = (u_0, \mathcal{A}v)_{0;S^+}.$$

Equation (2.9) can now be written in the form

$$p^2 B u_0 + \mathcal{A}u_0 = q. \tag{2.11}$$

Applying $\mathcal{A}^{-1}$ on both sides in (2.11), we arrive at the equivalent equation

$$p^2 \mathcal{A}^{-1} B u_0 + u_0 = \mathcal{A}^{-1} q \tag{2.12}$$

in the Banach space $\mathring{H}_{1,p}(S^+)$. We denote by $\mathcal{B}_0$ the restriction of $\mathcal{A}^{-1}B$ from $H_{-1}(S^+)$ to $\mathring{H}_1(S^+)$ and claim that $\mathcal{B}_0$ is compact on $\mathring{H}_1(S^+)$. Let $\{u_n\}_{n=1}^{\infty}$ be a weakly convergent sequence in $\mathring{H}_1(S^+)$. Since S^+ is a bounded domain, Rellich's lemma implies that the set $\{u_n\}_{n=1}^{\infty}$ is strongly compact in $L^2(S^+)$. Therefore, there is a subsequence $\{u_{n_j}\}_{j=1}^{\infty}$ that converges strongly in $L^2(S^+)$, consequently, also in $H_{-1}(S^+)$. Because $\mathcal{A}^{-1}B$ is continuous from $H_{-1}(S^+)$ to $\mathring{H}_1(S^+)$, the sequence $\{\mathcal{B}_0 u_{n_j}\}_{j=1}^{\infty}$ is strongly convergent in the space $\mathring{H}_1(S^+)$, which proves that $\mathcal{B}_0$ is a compact operator.

While studying the solvability of (2.12), we are under the conditions of the Fredholm Alternative, so (2.12)—hence, also (2.11)—has a unique solution $u_0 \in \mathring{H}_1(S^+)$ if and only if the homogeneous equation

$$p^2 \mathcal{A}^{-1} B u_0 + u_0 = 0 \tag{2.13}$$

has only the trivial solution. In turn, (2.13) can be rewritten in the equivalent form

$$p^2 B u_0 + \mathcal{A} u_0 = 0,$$

or

$$p^2 (B^{1/2} u_0, B^{1/2} v)_{0;S^+} + a_+(u_0, v) = 0 \quad \forall v \in \mathring{H}_1(S^+). \tag{2.14}$$

Taking $v = u_0$ in (2.14), writing $p = \sigma + i\tau$, and separating the real and imaginary parts, we arrive at

$$(\sigma^2 - \tau^2)\|B^{1/2} u_0\|_{0;S^+}^2 + a_+(u_0, u_0) = 0, \tag{2.15}$$

$$2\sigma\tau\|B^{1/2} u_0\|_{0;S^+}^2 = 0. \tag{2.16}$$

If $\tau \neq 0$, then from (2.16) it follows that $B^{1/2} u_0 = 0$; hence, $u_0 = 0$. If $\tau = 0$, then the equality $u_0 = 0$ follows from (2.15). So (2.11) is uniquely solvable in $\mathring{H}_1(S^+)$ for any $q \in H_{-1}(S^+)$.

We now establish estimate (2.8). Taking $v = u_0$ in (2.9) and separating the real and imaginary parts, we obtain

$$(\sigma^2 - \tau^2)\|B^{1/2} u_0\|_{0;S^+}^2 + a_+(u_0, u_0) = \operatorname{Re}(q, u_0)_{0;S^+}, \tag{2.17}$$

$$2\sigma\tau\|B^{1/2} u_0\|_{0;S^+}^2 = \operatorname{Im}(q, u_0)_{0;S^+}. \tag{2.18}$$

Multiplying (2.18) by $\sigma^{-1}\tau$ and adding the new equality to (2.17), we find that

$$|p|^2\|B^{1/2} u_0\|_{0;S^+}^2 + a_+(u_0, u_0)$$
$$= \operatorname{Re}(q, u_0)_{0;S^+} + \sigma^{-1}\tau \operatorname{Im}(q, u_0)_{0;S^+}$$
$$= \sigma^{-1} \operatorname{Re}\{\bar{p}(q, u_0)_{0;S^+}\}.$$

Because $p \in \bar{\mathbb{C}}_\kappa$, it follows that $\sigma \geq \kappa$. Taking into account the inequality $a_+(u_0, u_0) \geq c\|u_0\|_1^2$, we obtain

$$\|u_0\|_{1,p;S^+}^2 \leq c|p||(q, u_0)_{0;S^+}|,$$

from which

$$\|u_0\|_{1,p;S^+} \leq c|p|\|q\|_{-1,p;S^+}.$$

We now return to the full problem (2.7). Let $w = l^+ f \in H_1(S^+)$. We recall that since the extension operator l^+ is continuous (uniformly with respect to $p \in \mathbb{C}$), we have

$$\|w\|_{1,p;S^+} \le c \|f\|_{1/2,p;\partial S}. \tag{2.19}$$

Representing u in the form $u = w + u_0$, we see that $u_0 \in \mathring{H}_{1,p}(S^+)$ satisfies the equation

$$\begin{aligned}
p^2 (B^{1/2} u_0, B^{1/2} v) &+ a_+(u_0, v) \\
&= (q, v)_{0;S^+} - p^2 (B^{1/2} w, B^{1/2} v)_{0;S^+} - a_+(w, v) \\
&\qquad\qquad\qquad\qquad \forall v \in \mathring{H}_{1,p}(S^+). \tag{2.20}
\end{aligned}$$

We claim that the form $p^2 (B^{1/2} w, B^{1/2} v)_{0;S^+} + a_+(w, v)$ defines a bounded antilinear (conjugate linear) functional on $\mathring{H}_{1,p}(S^+)$; true,

$$\begin{aligned}
|p^2 (B^{1/2} w, &B^{1/2} v)_{0;S^+} + a_+(w, v)| \\
&\le c(|p|^2 \|w\|_{0;S^+} \|v\|_{0;S^+} + \|w\|_{1;S^+} \|v\|_1) \\
&\le c \|w\|_{1,p;S^+} \|v\|_{1,p} \quad \forall v \in \mathring{H}_{1,p}(S^+).
\end{aligned}$$

Consequently, there is $\tilde{q} \in H_{-1,p}(S^+)$ such that

$$p^2 (B^{1/2} w, B^{1/2} v)_{0;S^+} + a_+(w, v) = (\tilde{q}, v)_{0;S^+} \quad \forall v \in \mathring{H}_{1,p}(S^+)$$

and

$$\|\tilde{q}\|_{-1,p;S^+} \le c \|w\|_{1,p;S^+} \le c \|f\|_{1/2,p;\partial S}.$$

Equation (2.20) takes the form

$$p^2 (B^{1/2} u_0, B^{1/2} v)_{0;S^+} + a_+(u_0, v) = (q - \tilde{q}, v)_{0;S^+} \quad \forall v \in \mathring{H}_{1,p}(S^+).$$

As we already know, the latter problem is uniquely solvable and its solution satisfies the estimate

$$\begin{aligned}
\|u_0\|_{1,p} &\le c|p|(\|q\|_{-1,p;S^+} + \|\tilde{q}\|_{-1,p;S^+}) \\
&\le c|p|(\|q\|_{-1,p;S^+} + \|f\|_{1/2,p;\partial S}). \tag{2.21}
\end{aligned}$$

Combining (2.21) and (2.19), we arrive at (2.8).

If u_1 and u_2 are two solutions of (2.7), then $u_0 = u_1 - u_2 \in \mathring{H}_{1,p}(S^+)$ is a solution of (2.14); hence, $u_0 = 0$, and the theorem is proved. $\square$

As was mentioned above, we cannot repeat this proof in the case of S^- because Rellich's lemma is not valid for an unbounded domain. Consequently, we need to modify our approach.

2.2 Theorem. *For any $f \in H_{1/2,p}(\partial S)$ and $q \in H_{-1,p}(S^-)$, $p \in \bar{\mathbb{C}}_\kappa$, $\kappa > 0$, problem (D_p^-) has a unique solution $u \in H_{1,p}(S^-)$ and*

$$\|u\|_{1,p;S^-} \leq c|p|(\|q\|_{-1,p;S^-} + \|f\|_{1/2,p;\partial S}).$$

Proof. Again, first we assume that $f = 0$. In this case we seek $u_0 \in H_{1,p}(S^-)$ such that

$$p^2(B^{1/2}u_0, B^{1/2}v)_{0;S^-} + a_-(u_0, v) = (q, v)_{0;S^-} \quad \forall v \in \mathring{H}_{1,p}(S^-). \quad (2.22)$$

To prove the unique solvability of (2.22), we consider an auxiliary variational problem that consists in finding $u_0 \in \mathring{H}_1(S^-)$ such that

$$\tfrac{1}{2}\kappa^2(B^{1/2}u_0, B^{1/2}v)_{0;S^-} + a_-(u_0, v) = (q, v)_{0;S^-} \quad \forall v \in \mathring{H}_1(S^-), \quad (2.23)$$

where $q \in H_{-1}(S^-)$ is prescribed. Repeating the proof of Lemma 2.2 in [7], we find that there is a constant $c > 0$ such that

$$a_-(u, u) + \|u\|_{0;S^-}^2 \geq c\|u\|_{1;S^-}^2 \quad \forall u \in H_1(S^-). \quad (2.24)$$

From (2.24) it follows that the form

$$a_{-,\kappa}(u, v) = \tfrac{1}{2}\kappa^2(B^{1/2}u, B^{1/2}v)_{0;S^-} + a_-(u, v),$$

which is continuous on $\left[\mathring{H}_1(S^-)\right]^2$, is coercive on this space. The Lax–Milgram lemma then implies that (2.23) has a unique solution $u_0 \in \mathring{H}_1(S^-)$ for any $q \in H_{-1}(S^-)$. On the other hand, for any $u_0 \in \mathring{H}_1(S^-)$, the form $a_{-,\kappa}(u_0, v)$ generates a bounded antilinear (conjugate linear) functional on $\mathring{H}_1(S^-)$; therefore, it can be written in the form (2.23). This enables us to define an operator $\mathcal{A}_\kappa$ through the equality

$$(\mathcal{A}_\kappa u_0, v)_{0;S^-} = a_{-,\kappa}(u_0, v) \quad \forall v \in \mathring{H}_1(S^-),$$

which is a homeomorphism from $\mathring{H}_1(S^-)$ to $H_{-1}(S^-)$. Equation (2.23) can be rewritten as $\mathcal{A}_\kappa u_0 = q$. In turn, (2.22) can be written in the form

$$\mathcal{A}_\kappa u_0 + \left(p^2 - \tfrac{1}{2}\kappa^2\right)Bu_0 = q. \quad (2.25)$$

Applying $\mathcal{A}_\kappa^{-1}$ on both sides in (2.25), we arrive at the equivalent equation

$$u_0 + \left(p^2 - \tfrac{1}{2}\kappa^2\right)\mathcal{A}_\kappa^{-1}Bu_0 = \mathcal{A}_\kappa^{-1}q. \quad (2.26)$$

Setting $B^{1/2}u_0 = u_b$, we again rewrite (2.26) in the equivalent form

$$u_b + \left(p^2 - \tfrac{1}{2}\kappa^2\right)B^{1/2}\mathcal{A}_\kappa^{-1}B^{1/2}u_b = B^{1/2}\mathcal{A}_\kappa^{-1}q. \quad (2.27)$$

Equation (2.27) is solvable in $\mathring{H}_1(S^-)$. If $u_b \in \mathring{H}_1(S^-)$ is its solution, then u_b is, at the same time, the solution of (2.27) in $L^2(S^-)$. Conversely, let $u_b \in L^2(S^-)$ be a solution of (2.27). Since

$$B^{1/2}A_\kappa^{-1}B^{1/2}u_b \in \mathring{H}_1(S^-), \quad B^{1/2}A_\kappa^{-1}q \in \mathring{H}_1(S^-),$$

it follows that $u_b \in \mathring{H}_1(S^-)$. This means that problem (2.27) in $\mathring{H}_1(S^-)$ is equivalent to itself in $L^2(S^-)$. We now study the properties of the restriction $\mathcal{B}_\kappa$ of $B^{1/2}A_\kappa^{-1}B^{1/2}$ from $H_1(S^-)$ to $L^2(S^-)$.

Let $q, \psi \in L^2(S^-)$ be arbitrary, and let

$$u_0 = A_\kappa^{-1}B^{1/2}q, \quad v = A_\kappa^{-1}B^{1/2}\psi.$$

From the definition of A_κ it follows that

$$a_{-,\kappa}(u_0, v) = (B^{1/2}q, v)_{0;S^-},$$

$$a_{-,\kappa}(v, u_0) = (B^{1/2}\psi, u_0)_{0;S^-};$$

hence,

$$(B^{1/2}q, v)_{0;S^-} = \overline{(B^{1/2}\psi, u_0)}_{0;S^-} = (u_0, B^{1/2}\psi)_{0;S^-}. \qquad (2.28)$$

In turn, (2.28) can be rewritten as

$$(A_\kappa^{-1}B^{1/2}q, B^{1/2}\psi)_{0;S_-} = (B^{1/2}q, A_\kappa^{-1}B^{1/2}\psi)_{0;S^-},$$

or

$$(\mathcal{B}_\kappa q, \psi)_{0;S^-} = (q, \mathcal{B}_\kappa \psi)_{0;S^-} \quad \forall q, \psi \in L^2(S^-). \qquad (2.29)$$

By (2.29), $\mathcal{B}_\kappa$ is a symmetric operator on $L^2(S^-)$; therefore, it is self-adjoint (as a symmetric operator defined on the whole of a Hilbert space [3]).

Once again, let $q \in L^2(S^-)$ and $u_0 = A_\kappa^{-1}B^{1/2}q$; then

$$(\mathcal{B}_\kappa q, q)_{0;S^-} = (A_\kappa^{-1}B^{1/2}q, B^{1/2}q)_{0;S^-}$$

$$= (u_0, A_\kappa u_0)_{0;S^-} = a_{-,\kappa}(u_0, u_0) \geq 0,$$

which means that $\mathcal{B}_\kappa$ is nonnegative. Since the spectrum of a self-adjoint nonnegative operator lies on the half-line $[0, \infty)$ in the complex plane, every point of $\mathbb{C} \setminus [0, \infty)$ is a regular point for $\mathcal{B}_\kappa$ [3]. All that remains to do now is to remark that for any $p \in \mathbb{C}_\kappa$,

$$\left(\tfrac{1}{2}\kappa^2 - p^2\right)^{-1} \notin [0, \infty).$$

This implies that (2.27) is uniquely solvable for any $q \in H_{-1}(S^-)$. Consequently, equation (2.22) has a unique solution $u_0 \in \mathring{H}_{1,p}(S^-)$ for any $q \in H_{-1,p}(S^-)$. To complete the proof, we repeat the last part of the proof of Theorem 2.1, replacing the extension operator l^+ by l^-. $\qquad \square$

2.3 Solvability of the Time-dependent Problems

We start with the variational version of problems (DD$^{\pm}$) for the nonhomogeneous equation of motion. The classical formulation of (DD^{+}) asks for a function $u \in C^2(G^+) \cap C^1(\bar{G}^+)$ such that

$$B(\partial_t^2 u)(X) + (Au)(X) = q(X), \quad X \in G^+,$$
$$u(x, 0+) = (\partial_t u)(x, 0+) = 0, \quad x \in S^+, \tag{2.30}$$
$$u^+(X) = f(X), \quad X \in \Gamma.$$

Multiplying the first equation (2.30) by v^*, where $v \in C_0^\infty(\bar{G}^+)$ is such that $v^+ = 0$, integrating the new equality over S^+ with respect to x and over $[0, \infty)$ with respect to t, and taking into account the initial data for u and the boundary value of v, we arrive at

$$\int\limits_0^\infty \left[a_+(u, v) - (B^{1/2}\partial_t u, B^{1/2}\partial_t v)_{0;S^+} \right] dt = \int\limits_0^\infty (q, v)_{0;S^+} \, dt. \tag{2.31}$$

Conversely, if $u \in C^2(G^+) \cap C^1(\bar{G}^+)$ satisfies (2.31) for any $v \in C_0^\infty(\bar{G}^+)$ such that $v^+ = 0$, $u(x, 0+) = 0$, $x \in S^+$, and $u^+ = f$, then, integrating by parts in (2.31), we find that u is the solution of (2.30). Equation (2.31) suggests that the variational problem (DD$^+$) should consist in finding $u \in H_{1,0,\kappa}(G^+)$ that satisfies

$$\int\limits_0^\infty \left[a_+(u, v) - (B^{1/2}\partial_t u, B^{1/2}\partial_t v)_{0;S^+} \right] dt$$

$$= \int\limits_0^\infty (q, v)_{0;S^+} \, dt \quad \forall v \in C_0^\infty(\bar{G}^+), \ v^+ = 0, \tag{2.32}$$

$$\gamma^+ u = f,$$

where q and f are prescribed. Similarly, the variational problem (DD$^-$) consists in finding $u \in H_{1,0,\kappa}(G^-)$ that satisfies

$$\int\limits_0^\infty \left[a_-(u, v) - (B^{1/2}\partial_t u, B^{1/2}\partial_t v)_{0;S^-} \right] dt$$

$$= \int\limits_0^\infty (q, v)_{0;S^-} \, dt, \quad \forall v \in C_0^\infty(\bar{G}^-), \ v^- = 0, \tag{2.33}$$

$$\gamma^- u = f.$$

2.3 Theorem. *For any* $q \in H_{-1,1,\kappa}(G^+)$ *and* $f \in H_{1/2,1,\kappa}(\Gamma)$, $\kappa > 0$, *problems* (2.32) *and* (2.33) *have unique solutions* $u \in H_{1,0,\kappa}(G^\pm)$. *If* $q \in H_{-1,k,\kappa}(G^\pm)$ *and* $f \in H_{1/2,k,\kappa}(\Gamma)$, $k \in \mathbb{R}$, *then* $u \in H_{1,k-1,\kappa}(G^\pm)$ *and*

$$\|u\|_{1,k-1,\kappa;G^\pm} \le c(\|q\|_{-1,k,\kappa;G^\pm} + \|f\|_{1/2,k,\kappa;\Gamma}). \tag{2.34}$$

Proof. We prove the assertion for (DD^+); the case of (DD^-) is treated similarly.

Let $\hat{u} \in H_{1,p}(S^+)$ be the (weak) solution of the problem

$$p^2 B\hat{u}(x,p) + (A\hat{u})(x,p) = \hat{q}(x,p), \quad x \in S^+,$$
$$\gamma^+ \hat{u}(x,p) = \hat{f}(x,p), \quad x \in \partial S, \tag{2.35}$$

obtained by applying the Laplace transformation in (DD^+) with a nonhomogeneous equation of motion. The existence of $\hat{u}$ was proved in the previous section. For simplicity, in what follows we write

$$\hat{u}(\cdot,p) = U(p), \quad \hat{q}(\cdot,p) = Q(p), \quad \hat{f}(\cdot,p) = F(p),$$

and regard U, Q, and F as functions from $\mathbb{C}_\kappa$ to $H_1(S^+)$, $H_{-1}(S^+)$, and $H_{1/2}(\partial S)$, respectively.

We claim that the inverse Laplace transform u of $\hat{u}$ belongs to the space $H_{1,k-1,\kappa}(G^+)$ if $q \in H_{-1,k,\kappa}(G^+)$ and $f \in H_{1/2,k,\kappa}(\Gamma)$. To show this, first we verify that U is holomorphic from $\mathbb{C}_\kappa$ to $H_1(S^+)$. Let $p_0 \in \mathbb{C}_\kappa$, and let $K_R(p_0)$ be a circle with center at p_0 and radius R (to be specified later), and such that $\bar{K}_R(p_0) \subset \mathbb{C}_\kappa$. We recall that the solution $U(p_0)$ of the problem

$$p_0^2 BU(p_0) + (AU)(p_0) = Q(p_0),$$
$$\gamma^+ U(p_0) = F(p_0)$$

satisfies the estimates

$$\|U(p_0)\|_{1;S^+} \le c|p_0|(\|Q(p_0)\|_{-1,p_0;S^+} + \|F(p_0)\|_{1/2,p_0;\partial S})$$
$$\le c(\|Q(p_0)\|_{-1;S^+} + \|F(p_0)\|_{1/2;\partial S}).$$

Rewriting (2.35) in the form

$$p_0^2 BU(p) + (AU)(p) = Q(p) - (p^2 - p_0^2)BU(p),$$
$$\gamma^+ U(p) = F(p), \tag{2.36}$$

we see that

$$\|U(p)\|_{1;S^+} \le c(\|Q(p)\|_{-1;S^+} + \|F(p)\|_{1/2;\partial S}$$
$$+ |p^2 - p_0^2|\|U(p)\|_{-1;S^+}).$$

Since $\|U(p)\|_{-1;S^+} \leq \|U(p)\|_{1;S^+}$, it follows that for p satisfying

$$c|p^2 - p_0^2| \leq \tfrac{1}{2} \tag{2.37}$$

we have

$$\|U(p)\|_{1;S^+} \leq c(\|Q(p)\|_{-1;S^+} + \|F(p)\|_{1/2;\partial S}). \tag{2.38}$$

We choose $R > 0$ so that $\bar{K}_R(p_0) \subset \mathbb{C}_\kappa$ and (2.37) holds for $p \in K_R(p_0)$. Since Q and F are holomorphic from $\mathbb{C}_\kappa$ to $H_{-1}(S^+)$ and $H_{1/2}(\partial S)$, respectively, they are bounded in these spaces for $p \in \bar{K}_R(p_0)$. Estimate (2.38) shows that U is also bounded in $H_1(S^+)$ for $p \in \bar{K}_R(p_0)$. By (2.36),

$$p_0^2 B[U(p) - U(p_0)] + A[U(p) - U(p_0)]$$
$$= Q(p) - Q(p_0) - (p^2 - p_0^2)BU(p),$$
$$\gamma^+ U(p) - \gamma^+ U(p_0) = F(p) - F(p_0);$$

hence, for $p \in K_R(p_0)$,

$$\|U(p) - U(p_0)\|_{1;S^+}$$
$$\leq c(\|Q(p) - Q(p_0)\|_{-1;S^+} + \|F(p) - F(p_0)\|_{1/2;\partial S}$$
$$+ |p^2 - p_0^2| \|BU(p)\|_{-1;S^+}).$$

As $U(p)$ is bounded in $H_1(S^+)$—hence, also in $H_{-1}(S^+)$—for $p \in K_R(p_0)$, it follows that

$$\lim_{p \to p_0} \|U(p) - U(p_0)\|_{1;S^+} = 0,$$

which means that U is continuous from $\mathbb{C}_\kappa$ to $H_1(S^+)$ at p_0. Finally, let $V \in H_1(S^+)$ be the solution of the problem

$$p_0^2 BV - AV = Q'(p_0) - 2p_0 U(p_0),$$
$$\gamma^+ V = F'(p_0);$$

then the function

$$W(p) = (p - p_0)^{-1}[U(p) - U(p_0)] - V \in H_1(S^+)$$

satisfies

$$p_0^2 BW(p) + (AW)(p)$$
$$= (p - p_0)^{-1}[Q(p) - Q(p_0)] - Q'(p_0)$$
$$- B[(p + p_0)U(p) - 2p_0 U(p_0)],$$
$$\gamma^+ W = (p - p_0)^{-1}[F(p) - F(p_0)] - F'(p_0).$$

Next,

$$\|W(p)\|_{1;S^+} \leq c \left\{ \left\| \frac{Q(p) - Q(p_0)}{p - p_0} - Q'(p_0) \right\|_{-1;S^+} \right.$$

$$+ \left\| \frac{F(p) - F(p_0)}{p - p_0} - F'(p_0) \right\|_{1/2;\partial S}$$

$$\left. + \|(p + p_0)U(p) - 2p_0 U(p_0)\|_{-1;S^+} \right\}.$$

Since U is continuous at p_0,

$$\lim_{p \to p_0} \|W(p)\|_{1;S^+} = 0,$$

which means that $U'(p_0)$ exists and $U'(p_0) = V$. The arbitrariness of p_0 in $\mathbb{C}_\kappa$ implies that the mapping U is holomorphic from $\mathbb{C}_\kappa$ to $H_1(S^+)$.

Given that

$$\|U(p)\|_{1,p;S^+} \leq c|p|(\|Q(p)\|_{-1,p;S^+} + \|F(p)\|_{1/2,p;\partial S}),$$

we have

$$\|u\|_{1,k-1,\kappa;G^+}^2 = \sup_{\sigma > \kappa} \int_{-\infty}^{\infty} (1 + |p|^2)^{k-1} \|U(p)\|_{1,p;S^+}^2 \, d\tau$$

$$\leq c \sup_{\sigma > \kappa} \int_{-\infty}^{\infty} (1 + |p|^2)^k (\|Q(p)\|_{-1,p;S^+}^2 + \|F(p)\|_{1/2,p;\partial S}^2) \, d\tau$$

$$\leq c(\|q\|_{-1,k,\kappa;G^+}^2 + \|f\|_{1/2,k,\kappa;\Gamma}^2), \tag{2.39}$$

where $p = \sigma + i\tau$. This confirms (2.34).

To complete the proof of the theorem, we need to check that u is the only solution of (2.32). We recall [11] that any two functions $f_1(t)$ and $f_2(t)$ such that

$$\int_0^{\infty} e^{-2\kappa_\nu t} |f_\nu(t)|^2 \, dt < \infty, \quad \nu = 1, 2,$$

satisfy Parseval's equality

$$\int_0^{\infty} e^{-(\kappa_1 + \kappa_2)t} f_1(t) \overline{f_2(t)} \, dt = \frac{1}{2\pi} \int_{-\infty}^{\infty} F_1(\kappa_1 + i\tau) \overline{F_2(\kappa_2 + i\tau)} \, d\tau, \tag{2.40}$$

where $F_\nu(p)$ are the Laplace transforms of $f_\nu(t)$. Let $v \in \mathring{C}_0^\infty(\bar{G}^+)$ be such that $\gamma^+ v = 0$. We make the notation

$$v(x,0) = v_0(x), \quad v_0 \in \mathring{H}_1(S^+), \quad v(\cdot,p) = V(p), \quad v_0(\cdot) = V_0,$$
$$p = \sigma + i\tau, \quad p^* = -\sigma + i\tau,$$

choose any $\sigma > \kappa$ and fix it, then take $\kappa_1 = \sigma$ and $\kappa_2 = -\sigma$ in (2.40) and find that the Laplace transform of $\partial_t v$ at the point p^* is $p^* V(p^*) - V_0$, and that

$$\int_0^\infty \left[a_+(u,v) - (B^{1/2}\partial_t u, B^{1/2}\partial_t v)_{0;S^+} - (q,v)_{0;S^+} \right] dt$$

$$= (2\pi)^{-1} \int_{-\infty}^\infty \left[a_+(U(p), V(p^*)) - (B^{1/2}pU(p), B^{1/2}(p^*V(p^*) - V_0))_{0;S^+} \right.$$

$$\left. - (Q(p), V(p^*))_{0;S^+} \right] d\tau. \qquad (2.41)$$

Since $U(p)$ is a weak solution of (2.4), it follows that

$$a_+(U(p), W) + p^2 (B^{1/2}U(p), B^{1/2}W)_{0;S^+}$$

$$= (Q(p), W)_{0;S^+} \quad \forall W \in \mathring{H}_1(S^+). \qquad (2.42)$$

Taking $W = V(p^*) - (p^*)^{-1}V_0$ in (2.42), we obtain

$$a_+(U(p), V(p^*)) - (B^{1/2}pU(p), B^{1/2}(p^*V(p^*) - V_0))_{0;S^+}$$

$$- (Q(p), V(p^*))_{0;S^+}$$

$$= a_+(U(p), (p^*)^{-1}V_0) - (Q(p), (p^*)^{-1}V_0)_{0;S^+}$$

$$= p^{-1}(Q(p), V_0)_{0;S^+} - p^{-1}a_+(U(p), V_0).$$

Therefore, (2.41) takes the form

$$\int_0^\infty \left[a_+(u,v) - (B^{1/2}\partial_t u, B^{1/2}\partial_t v)_{0;S^+} - (q,v)_{0;S^+} \right] dt$$

$$= (2\pi)^{-1} \int_{-\infty}^\infty p^{-1} \left[(Q(p), V_0)_{0;S^+} - a_+(U(p), V_0) \right] d\tau. \qquad (2.43)$$

We claim that the right-hand side in (2.43) vanishes. First, we remark that

$$\|\hat{q}\|^2_{-1;S^+} \le (1 + |p|^2)\|\hat{q}\|^2_{-1,p;S^+} \quad \forall \hat{q} \in H_{-1}(S^+). \qquad (2.44)$$

For if $\hat{q} \in H_{-1}(\mathbb{R}^2)$ and $\tilde{q}$ is its Fourier transform, then

$$\|\hat{q}\|_{-1}^2 = \int_{\mathbb{R}^2} (1+|\xi|^2)^{-1}|\tilde{q}(\xi)|^2 \, d\xi$$

$$\leq (1+|p|^2) \int_{\mathbb{R}^2} (1+|p|^2)^{-1}(1+|\xi|^2)^{-1}|\tilde{q}(\xi)|^2 \, d\xi$$

$$\leq (1+|p|^2) \int_{\mathbb{R}^2} (1+|p|^2+|\xi|^2)^{-1}|\tilde{q}(\xi)|^2 \, d\xi$$

$$= (1+|p|^2)\|\hat{q}\|_{-1,p}^2. \tag{2.45}$$

Inequality (2.44) follows from (2.45) and the definition of the norms on the spaces $H_{-1}(S^+)$ and $H_{-1,p}(S^+)$. By (2.44),

$$\int_{-\infty}^{\infty} |(Q(p), V_0)_{0;S^+}|^2 \, d\tau \leq \|V_0\|_1^2 \int_{-\infty}^{\infty} \|Q(p)\|_{-1;S^+}^2 \, d\tau$$

$$\leq \|V_0\|_1^2 \int_{-\infty}^{\infty} (1+|p|^2)\|Q(p)\|_{-1,p;S^+}^2 \, d\tau$$

$$\leq \|V_0\|_1^2 \|q\|_{-1,1,\kappa;G^+}^2 < \infty.$$

Consequently, the function $\varphi(t) = (q, v_0)_{0;S^+}$ satisfies

$$\int_0^{\infty} e^{-2\sigma t}|\varphi(t)|^2 \, dt < \infty,$$

so $\varphi \in L^1_{\text{loc}}(0, \infty)$, which implies that

$$\psi(t) = \int_0^t \varphi(\lambda) \, d\lambda$$

is continuous for $t \in [0, \infty)$; in particular, $\psi(0) = 0$. We have

$$0 = \psi(0) = (2\pi)^{-1} \int_{-\infty}^{\infty} p^{-1}(Q(p), V_0)_{0;S^+} \, d\tau.$$

Next,

$$\int_{-\infty}^{\infty} |a_+(U(p), V_0)|^2 \, d\tau \leq c\|V_0\|_1^2 \int_{-\infty}^{\infty} \|U(p)\|_{1,p;S^+}^2 \, d\tau$$

$$\leq c\|V_0\|_1^2 \|u\|_{1,0,\kappa;G^+}^2,$$

and the above arguments yield

$$\int_{-\infty}^{\infty} p^{-1} a_+(U(p), V_0) \, d\tau = 0.$$

Equality (2.43) now leads to

$$\int_0^{\infty} \left[a_+(u, v) - (B^{1/2}\partial_t u, B^{1/2}\partial_t v)_{0;S^+} \right] dt = \int_0^{\infty} (q, v)_{0;S^+} \, dt,$$

which means that u is a weak solution of (2.32). To prove that this solution is unique, suppose that $u \in H_{1,0,\kappa}(G^+)$, $\gamma^+ u = 0$, satisfies

$$\int_0^{\infty} \left[a_+(u, v) - (B^{1/2}\partial_t u, B^{1/2}\partial_t v)_{0;S^+} \right] dt = 0$$

$$\forall v \in C_0^{\infty}(\bar{G}^+), \ \gamma^+ v = 0. \qquad (2.46)$$

We fix an arbitrary $T > 0$. It is obvious that

$$u \in H_1((0,T); L^2(S^+)) \cap L^2((0,T); \mathring{H}_1(S^+));$$

that is, $U(t) = u(\cdot, t)$, regarded as a vector-valued function from $(0, T)$ to $L^2(S^+)$, belongs to $H_1(0, T)$ and

$$\int_0^T \left\{ \|U(t)\|_{0;S^+}^2 + \|U'(t)\|_{0;S^+}^2 \right\} dt < \infty. \qquad (2.47)$$

The same function, regarded as a mapping from $(0, T)$ to $\mathring{H}_1(S^+)$, belongs to $L^2(0, T)$ and

$$\int_0^T \|U(t)\|_1^2 \, dt < \infty. \qquad (2.48)$$

From (2.47) and (2.48) we see that $H_1((0,T); L^2(S^+)) \cap L^2((0,T); \mathring{H}_1(S^+))$ can be equipped with the norm

$$\|u\|_{1;G_T^+}^2 = \int\limits_0^T \int\limits_{S^+} \sum_{|\alpha|+\alpha_t \leq 1} |\partial^{\alpha+\alpha_t} u(x,t)|^2 \, dx \, dt,$$

where $G_T^+ = S^+ \times (0,T)$.

We now construct the function $Z(t) = z(\cdot, t)$, where

$$z(x,t) = \begin{cases} -\int\limits_t^T u(x,\tau) \, d\tau, & t \leq T, \\ 0, & t > T. \end{cases} \tag{2.49}$$

Clearly, the restriction of Z to $(0,T)$ belongs to $H_1((0,T); L^2(S^+)) \cap L^2((0,T); \mathring{H}_1(S^+))$ and that z can be approximated with any accuracy in the norm $\|\cdot\|_{1;G_T^+}$ by means of elements $v \in C_0^\infty(\bar{G}^+)$ such that $\gamma^+ v = 0$. Hence, we may set $v = z$ in (2.46) and obtain

$$\int\limits_0^T \left[a_+(u,z) - (B^{1/2}\partial_t u, B^{1/2}\partial_t z)_{0;S^+} \right] dt = 0.$$

We remark that

$$Z'(t) = (\partial_t z)(\cdot, t) = \begin{cases} u(\cdot, t), & t < T, \\ 0, & t > T, \end{cases} \tag{2.50}$$

and rewrite (2.46) in the form

$$\int\limits_0^T \left[a_+(\partial_t z, z) - (B^{1/2}\partial_t u, B^{1/2}u)_{0;S^+} \right] dt = 0,$$

or

$$\int\limits_0^T \frac{d}{dt} \left[a_+(z,z) - \|B^{1/2}u\|_{0;S^+}^2 \right] dt = 0. \tag{2.51}$$

Since U, regarded as a mapping from $(0,T)$ to $L^2(S^+)$, belongs to $H_1(0,T)$, it is absolutely continuous on $[0,T]$; hence,

$$\int\limits_0^T \frac{d}{dt} \|B^{1/2}u\|_{0;S^+}^2 \, dt = \|B^{1/2}U(T)\|_{0;S^+}^2 - \|B^{1/2}U(0)\|_{0;S^+}^2$$

$$= \|B^{1/2}U(T)\|_{0;S^+}^2. \tag{2.52}$$

From (2.49) and (2.50), it follows that Z, as a mapping from $(0, T)$ to $\overset{\circ}{H}_1(S^+)$, belongs to $H_1(0, T)$; hence, Z is absolutely continuous on $[0, T]$ and

$$\int_0^T \frac{d}{dt} a_+(z, z) \, dt = a_+(Z(T), Z(T)) - a_+(Z(0), Z(0))$$

$$= -a_+(Z(0), Z(0)). \tag{2.53}$$

Formulas (2.51)–(2.53) now imply that

$$a_+(Z(0), Z(0)) + \|B^{1/2}U(T)\|_{0;S^+}^2 = 0;$$

therefore, $u(T) = u(\cdot, T) = 0$ for any $T > 0$, which completes the proof. $\square$

3

Problems with Neumann Boundary Conditions

3.1 The Poincaré–Steklov Operators

In this section we study the properties of the dynamic analogs of the Poincaré–Steklov operators introduced in [7]. We begin by considering these operators in Sobolev spaces with a parameter.

Let $f \in H_{1/2,p}(\partial S)$, $p \in \mathbb{C}_0$, and let $u \in H_{1,p}(S^{\pm})$ be the (unique) solutions of the problems

$$p^2(B^{1/2}u, B^{1/2}v)_{0;S^{\pm}} + a_{\pm}(u, v) = 0 \quad \forall v \in \mathring{H}_{1,p}(S^{\pm}),$$

$$\gamma^{\pm}u = f.$$

$$(3.1)$$

Also, let φ be an arbitrary element of $H_{1/2,p}(\partial S)$, and let $w \in H_{1,p}(S^{\pm})$ be such that $\gamma^{\pm}w = \varphi$. For every $p \in \mathbb{C}_0$, we define a pair of Poincaré–Steklov operators $\mathcal{T}_p^{\pm}$ on $H_{1/2,p}(\partial S)$ by means of the equality

$$(\mathcal{T}_p^{\pm}f, \varphi)_{0;\partial S} = \pm\big[p^2(B^{1/2}u, B^{1/2}w)_{0;S^{\pm}} + a_{\pm}(u, w)\big].$$

$$(3.2)$$

It is obvious that (3.2) defines $\mathcal{T}_p^{\pm}$ correctly. For if $w_1, w_2 \in H_{1,p}(S^{\pm})$ are such that

$$\gamma^{\pm}w_1 = \gamma^{\pm}w_2 = \varphi,$$

then $v = w_1 - w_2 \in \mathring{H}_{1,p}(S^{\pm})$ and, by (3.1),

$$p^2(B^{1/2}u, B^{1/2}w_1)_{0;S^{\pm}} + a_{\pm}(u, w_1)$$

$$= p^2\big(B^{1/2}u, B^{1/2}(v + w_2)\big)_{0;S^{\pm}} + a_{\pm}(u, v + w_2)$$

$$= p^2(B^{1/2}u, B^{1/2}w_2)_{0;S^{\pm}} + a_{\pm}(u, w_2).$$

3.1 Lemma. *For any $p \in \mathbb{C}_0$, the operators $\mathcal{T}_p^\pm$ are homeomorphisms from $H_{1/2,p}(\partial S)$ to $H_{-1/2,p}(\partial S)$, and for any $f \in H_{1/2,p}(\partial S)$, $p \in \bar{\mathbb{C}}_\kappa$, $\kappa > 0$,*

$$\|\mathcal{T}_p^\pm f\|_{-1/2,p;\partial S} \leq c|p|\|f\|_{1/2,p;\partial S}, \tag{3.3}$$

$$\|f\|_{1/2,p;\partial S} \leq c|p|\|\mathcal{T}_p^\pm f\|_{-1/2,p;\partial S}. \tag{3.4}$$

Proof. Let $f, \varphi \in H_{1/2,p}(\partial S)$, let $w = l^\pm \varphi$, and let $u \in H_{1,p}(S^\pm)$ be the solutions of (3.1) in $S^\pm$. By (3.2),

$$|(\mathcal{T}_p^\pm f, \varphi)_{0;\partial S}|^2 \leq c\left[\|u\|_{1,p;S^\pm}^2 \|v\|_{1,p;S^\pm}^2 + a_\pm(u,u)a_\pm(w,w)\right]$$
$$\leq c\|u\|_{1,p;S^\pm}^2 \|\varphi\|_{1/2,p;\partial S}^2;$$

therefore, $\mathcal{T}_p^\pm f \in H_{-1/2,p}(\partial S)$ and

$$\|\mathcal{T}_p^\pm f\|_{-1/2,p;\partial S} \leq c\|u\|_{1,p;S^\pm}. \tag{3.5}$$

Taking $\varphi = f$ and $w = u$ in (3.1), we obtain

$$p^2\|B^{1/2}u\|_{0;S^\pm}^2 + a_\pm(u,u) = \pm(\mathcal{T}_p^\pm f, f)_{0;\partial S},$$

where $p = \sigma + i\tau$. Next, we separate the real and imaginary parts above to find that

$$(\sigma^2 - \tau^2)\|B^{1/2}u\|_{0;S^\pm}^2 + a_\pm(u,u) = \pm\operatorname{Re}(\mathcal{T}_p^\pm f, f)_{0;\partial S}, \tag{3.6}$$

$$2\sigma\tau\|B^{1/2}u\|_{0;S^\pm}^2 = \pm\operatorname{Im}(\mathcal{T}_p^\pm f, f)_{0;\partial S}. \tag{3.7}$$

Multiplying (3.7) by $\sigma^{-1}\tau$ and adding the result to (3.6), we arrive at the equalities

$$|p|^2\|B^{1/2}u\|_{0;S^\pm}^2 + a_\pm(u,u) = \pm\sigma^{-1}\operatorname{Re}\left\{\bar{p}(\mathcal{T}_p^\pm f, f)_{0;\partial S}\right\}. \tag{3.8}$$

Since, as shown in [7],

$$a_\pm(u,u) + \|u\|_{0;S^\pm}^2 \geq c\|u\|_{1;S^\pm}^2,$$

we easily deduce from (3.8) that

$$\|u\|_{1,p;S^\pm}^2 \leq \pm c\sigma^{-3}\operatorname{Re}\left\{\bar{p}(\mathcal{T}_p^\pm f, f)_{0;\partial S}\right\}; \tag{3.9}$$

hence, for $p \in \bar{\mathbb{C}}_\kappa$,

$$\|u\|_{1,p;S^\pm}^2 \leq c|p|\|\mathcal{T}_p^\pm f\|_{-1/2,p;\partial S}\|f\|_{1/2,p;\partial S}, \tag{3.10}$$

and the trace theorem implies that

$$\|u\|_{1,p;S^{\pm}} \leq c|p|\|\mathcal{T}_p^{\pm}f\|_{-1/2,p;\partial S}. \tag{3.11}$$

From (3.5) and (3.10) it follows that

$$\|\mathcal{T}_p^{\pm}f\|^2_{-1/2,p;\partial S} \leq c|p|\|\mathcal{T}_p^{\pm}f\|_{-1/2,p;\partial S}\|f\|_{1/2,p;\partial S};$$

therefore,

$$\|\mathcal{T}_p^{\pm}f\|_{-1/2,p;\partial S} \leq c|p|\|f\|_{1/2,p;\partial S}.$$

This proves the continuity of $\mathcal{T}_p^{\pm}$ from $H_{1/2,p}(\partial S)$ to $H_{-1/2,p}(\partial S)$, and (3.3).

We now show that the inverse operators $(\mathcal{T}_p^{\pm})^{-1}$ exist. By the trace theorem and (3.11),

$$\|f\|_{1/2,p;\partial S} \leq c\|u\|_{1,p;S^{\pm}} \leq c|p|\|\mathcal{T}_p^{\pm}f\|_{-1/2,p;\partial S},$$

so $(\mathcal{T}_p^{\pm})^{-1}$ are continuous from $H_{-1/2,p;\partial S}$ to $H_{1/2,p}(\partial S)$ and (3.4) holds.

Suppose that the ranges of $\mathcal{T}_p^{\pm}$ are not dense in $H_{-1/2,p}(\partial S)$. Then there is a nonzero $\varphi \in H_{1/2,p}(\partial S)$ such that $(\mathcal{T}_p^{\pm}f, \varphi)_{0;\partial S} = 0$ for all $f \in H_{1/2,p}(\partial S)$. Taking $f = \varphi$, we see that $(\mathcal{T}_p^{\pm}\varphi, \varphi)_{0;\partial S} = 0$. If z is the solution of either of the problems (3.1) with boundary value φ, then (3.8) implies that $z = 0$; consequently, $\varphi = 0$. This contradiction completes the proof. □

At this stage, we can define operators $\hat{\mathcal{T}}^{\pm}$ and $(\hat{\mathcal{T}}^{\pm})^{-1}$ on the elements $F(p) = \hat{f}(\,\cdot\,, p)$ and $G(p) = \hat{g}(\,\cdot\,, p)$ of $H_{1/2,k,\kappa}(\partial S)$ and $H_{-1/2,k,\kappa}(\partial S)$, $k \in \mathbb{R}$, respectively, by setting

$$(\hat{\mathcal{T}}^{\pm}F)(p) = (\mathcal{T}_p^{\pm}\hat{f})(\,\cdot\,, p),$$

$$((\hat{\mathcal{T}}^{\pm})^{-1}G)(p) = ((\mathcal{T}_p^{\pm})^{-1}\hat{g})(\,\cdot\,, p).$$

Finally, we return to the spaces of originals and define the Poincaré–Steklov operators $\mathcal{T}^{\pm}$ and $(\mathcal{T}^{\pm})^{-1}$ on the elements $f \in H_{1/2,k,\kappa}(\Gamma)$ and $g \in H_{-1/2,k,\kappa}(\Gamma)$, $k \in \mathbb{R}$, by means of the equalities

$$\mathcal{T}^{\pm}f = \mathcal{L}^{-1}\hat{\mathcal{T}}^{\pm}\mathcal{L}f,$$

$$(\mathcal{T}^{\pm})^{-1}g = \mathcal{L}^{-1}(\hat{\mathcal{T}}^{\pm})^{-1}\mathcal{L}g.$$

3.2 Theorem. *For any $\kappa > 0$ and $k \in \mathbb{R}$, the operators $\mathcal{T}^{\pm}$ are continuous and injective from $H_{1/2,k,\kappa}(\Gamma)$ to $H_{-1/2,k-1,\kappa}(\Gamma)$, and their ranges are dense in $H_{-1/2,k-1,\kappa}(\Gamma)$. Their inverses $(\mathcal{T}^{\pm})^{-1}$, extended by continuity from the ranges of $\mathcal{T}^{\pm}$ to $H_{-1/2,k,\kappa}(\Gamma)$, are continuous and injective from $H_{-1/2,k,\kappa}(\Gamma)$ to $H_{1/2,k-1,\kappa}(\Gamma)$ for any $k \in \mathbb{R}$, and their ranges are dense in the corresponding spaces.*

Proof. Let

$$f \in H_{1/2,k,\kappa}(\Gamma), \quad g \in H_{-1/2,k,\kappa}(\Gamma),$$
$$\mathcal{L}f(x,t) = \hat{f}(x,p), \quad \mathcal{L}g(x,t) = \hat{g}(x,p).$$

First, we show that if $F(p) = \hat{f}(\cdot,p)$ is holomorphic from $\mathbb{C}_\kappa$ to $H_{1/2}(\partial S)$, then $\Phi(p) = (\mathcal{T}_p^\pm \hat{f})(\cdot,p)$ is holomorphic from $\mathbb{C}_\kappa$ to the space $H_{-1/2}(\partial S)$. In the proof of Theorem 2.3 we established that if $F(p)$ is holomorphic from $\mathbb{C}_\kappa$ to $H_{1/2}(\partial S)$, then $U(p) = \hat{u}(\cdot,p)$, where $\hat{u}(x,p)$ is the solution of either of the problems (3.1) with boundary data $\hat{f}(x,p)$, is holomorphic from $\mathbb{C}_\kappa$ to $H_1(S^\pm)$. We now take any $\varphi \in H_{1/2}(\partial S)$ and construct $w = l^\pm \varphi$. Let $p_0 \in \mathbb{C}_\kappa$. By (3.2),

$$\left(\Phi(p) - \Phi(p_0), \varphi\right)_{0;\partial S}$$
$$= \pm\left[(p^2 B^{1/2} U(p) - p_0^2 B^{1/2} U(p_0), B^{1/2} w)_{0;S^\pm}\right.$$
$$\left. + a_\pm\left(U(p) - U(p_0), w\right)\right]$$
$$= \pm\left[(p^2 - p_0^2)(B^{1/2} U(p), B^{1/2} w)_{0;S^\pm}\right.$$
$$+ p_0^2\left(B^{1/2}(U(p) - U(p_0)), B^{1/2} w\right)_{0;S^\pm}$$
$$\left. + a_\pm\left(U(p) - U(p_0), w\right)\right];$$

therefore,

$$\left(\frac{\Phi(p) - \Phi(p_0)}{p - p_0}, \varphi\right)_{0;\partial S}$$
$$= \pm\left[(p + p_0)(B^{1/2} U(p), B^{1/2} w)_{0;S^\pm}\right.$$
$$\left. + p_0^2\left(B^{1/2}\frac{U(p) - U(p_0)}{p - p_0}, B^{1/2} w\right)_{0;S^\pm} + a_\pm\left(\frac{U(p) - U(p_0)}{p - p_0}, w\right)\right].$$

Let $K_R(p_0)$ be a circle with the center at p_0 and radius R such that $\bar{K}_R(p_0) \subset \mathbb{C}_\kappa$. Since

$$\|\hat{u}\|_{1,p;S^\pm} \le c\|\hat{u}\|_{1;S^\pm}, \quad p \in \bar{K}_R(p_0),$$

where $c = \text{const} > 0$ is independent of p, it follows that

$$\lim_{p \to p_0}\left(\frac{\Phi(p) - \Phi(p_0)}{p - p_0}, \varphi\right)_{0;\partial S}$$
$$= \pm\left[2p_0(B^{1/2} U(p_0), B^{1/2} w)_{0;S^\pm}\right.$$
$$\left. + p^2(B^{1/2} U'(p_0), B^{1/2} w)_{0;S^\pm} + a_\pm(U'(p_0), w)\right].$$

This shows that $\Phi(p)$ is weakly holomorphic from $\mathbb{C}_\kappa$ to $H_{-1/2}(\partial S)$. According to Dunford's theorem [18], any weakly holomorphic mapping is holomorphic in the strong sense, so our assertion is proved.

Next, by (3.3) with $p = \sigma + i\tau$ and the definition of $\mathcal{T}^\pm$,

$$\|\mathcal{T}^\pm f\|^2_{-1/2,k-1,\kappa;\Gamma}$$

$$= \sup_{\sigma>\kappa} \int_{-\infty}^{\infty} (1 + |p|^2)^{k-1}\|(\mathcal{T}_p^\pm F)(p)\|^2_{-1/2,p;\partial S}\, d\tau$$

$$\leq c \sup_{\sigma>\kappa} \int_{-\infty}^{\infty} (1 + |p|^2)^{k}\|F(p)\|^2_{1/2,p;\partial S}\, d\tau$$

$$= c\|f\|^2_{1/2,k,\kappa;\Gamma}.$$

We now go over to the case of $(\mathcal{T}^\pm)^{-1}$. For any $p \in \mathbb{C}_0$, the Neumann boundary value problems

$$Bp^2\hat{u}(x,p) + (A\hat{u})(x,p) = 0, \quad x \in S^\pm,$$

$$(T\hat{u})^\pm(x,p) = \hat{g}(x,p), \quad x \in \partial S,$$

admit a variational version that consists in finding $\hat{u} \in H_{1,p}(S^\pm)$ such that

$$p^2(B^{1/2}\hat{u}, B^{1/2}v)_{0;S^\pm} + a_\pm(\hat{u}, v) = \pm(\hat{g}, \gamma^\pm v)_{0;\partial S} \quad \forall v \in H_{1,p}(S^\pm), \quad (3.12)$$

where $\hat{g} \in H_{-1/2,p}(\partial S)$ is prescribed. Let

$$\hat{f} = (\mathcal{T}_p^\pm)^{-1}\hat{g} \in H_{1/2,p}(\partial S),$$

and let $\hat{u} \in H_{1,p}(S^\pm)$ be the solutions of problems (3.1). Since $\hat{g} = \mathcal{T}_p^\pm\hat{f}$, from (3.2) it follows that $\hat{u}$ are the (unique) solutions of problems (3.12). In addition, (3.11) implies that

$$\|\hat{u}\|_{1,p;S^\pm} \leq c|p|\|\hat{g}\|_{-1/2,p;\partial S}.$$

Taking this inequality into account and repeating the proof of Theorem 2.3, we conclude that if $G(p) = \hat{g}(\cdot, p)$ is holomorphic from $\mathbb{C}_\kappa$ to $H_{-1/2}(\partial S)$, then the solutions $U(p) = \hat{u}(\cdot, p)$ of (3.12) are holomorphic from $\mathbb{C}_\kappa$ to $H_1(S^\pm)$, respectively. Since the trace operators $\gamma^\pm$ are continuous from $H_1(S^\pm)$ to $H_{1/2}(\partial S)$, we deduce that $F(p) = ((\mathcal{T}_p^\pm)^{-1}G)(p) = \gamma^\pm U(p)$ are holomorphic from $\mathbb{C}_\kappa$ to $H_{1/2}(\partial S)$. Also, from (3.4) it follows that $(\mathcal{T}^\pm)^{-1}$ are continuous from $H_{-1/2,k,\kappa}(\Gamma)$ to $H_{1/2,k-1,\kappa}(\Gamma)$ for any $k \in \mathbb{R}$.

To complete the proof, we only need to show that the ranges of $\mathcal{T}^\pm$ and $(\mathcal{T}^\pm)^{-1}$ are dense in the corresponding spaces. Let $\mathcal{H}_k^\pm$ be the ranges of the operators

$$\mathcal{T}^\pm : H_{1/2,k,\kappa}(\Gamma) \to H_{-1/2,k-1,\kappa}(\Gamma).$$

We take any $g \in H_{-1/2,k+1,\kappa}(\Gamma)$ and set $f = (\mathcal{T}^\pm)^{-1}g \in H_{1/2,k,\kappa}(\Gamma)$. Since $g = \mathcal{T}^\pm(\mathcal{T}^\pm)^{-1}g$, we have $H_{-1/2,k+1,\kappa}(\Gamma) \subset \mathcal{H}_k^\pm$. It is obvious that for any $k \in \mathbb{R}$, the space $H_{-1/2,k+1,\kappa}(\Gamma)$ is dense in $H_{-1/2,k-1,\kappa}(\Gamma)$; hence, $\mathcal{H}_k^\pm$ are dense in $H_{-1/2,k-1,\kappa}(\Gamma)$. The case of $(\mathcal{T}^\pm)^{-1}$ is treated similarly. $\square$

3.2 Solvability of the Problems

Consider problems (DN$^\pm$), whose classical formulation was given in §1.1. Their variational versions consist in finding $u \in H_{1,0,\kappa}(G^\pm)$ such that, for g prescribed on Γ,

$$\int_0^\infty \left[a_\pm(u,v) - (B^{1/2}\partial_t u, B^{1/2}\partial_t v)_{0;S^\pm} \right] dt$$

$$= \pm \int_0^\infty (g, v^\pm)_{0;\partial S}\, dt \quad \forall v \in C_0^\infty(\bar{G}^\pm). \qquad (3.13)$$

3.3 Theorem. *For any* $g \in H_{-1/2,1,\kappa}(\Gamma)$, $\kappa > 0$, *problems* (3.13) *have unique solutions* $u \in H_{1,0,\kappa}(G^\pm)$. *If* $g \in H_{-1/2,k,\kappa}(\Gamma)$, $k \in \mathbb{R}$, *then these solutions belong to* $H_{1,k-1,\kappa}(G^\pm)$, *respectively, and*

$$\|u\|_{1,k-1,\kappa;G^\pm} \leq c\|g\|_{-1/2,k,\kappa;\Gamma}. \qquad (3.14)$$

Proof. Let $\hat{u}(x,p)$, $p \in \bar{\mathbb{C}}_\kappa$, be the solutions of problems (3.12). It has already been shown that $U(p) = \hat{u}(\cdot,p)$ are holomorphic from $\mathbb{C}_\kappa$ to $H_1(S^\pm)$. This and (3.11) imply that $u = \mathcal{L}^{-1}U \in H_{1,k-1,\kappa}(G^\pm)$, and that (3.14) holds. The proof that u satisfies (3.13) is similar to that of the analogous assertion for (DD$^\pm$). Consequently, we only need to check the uniqueness of the solution to (3.13).

Let u_1 and u_2 be two solutions of (3.13). Then $u = u_1 - u_2 \in H_{1,0,\kappa}(G^\pm)$ satisfies

$$\int_0^\infty \left[a_\pm(u,v) - (B^{1/2}\partial_t u, B^{1/2}\partial_t v)_{0;S^\pm} \right] dt = 0 \quad \forall v \in C_0^\infty(\bar{G}^\pm).$$

Repeating the proof of Theorem 2.3 with $H_1(S^\pm)$ instead of $\mathring{H}_1(S^\pm)$, we conclude that $u = 0$, as required. $\square$

The more general problems (DN$^\pm$) for the nonhomogeneous equation of motion can be reduced to the above by means of an appropriate substitution (see Chapter 9).

4

Boundary Integral Equations for Problems with Dirichlet and Neumann Boundary Conditions

4.1 Time-dependent Potentials

We recall that the single-layer (retarded) potential $V\alpha$ of density α defined on $\partial S \times \mathbb{R}$ was introduced in §1.3 as

$$(V\alpha)(X) = \int_{-\infty}^{\infty} \int_{\partial S} D(x - y, t - \tau)\alpha(y, \tau)\, ds_y\, d\tau.$$

If $\alpha(X) = 0$ for $t < 0$, then the Laplace transform of this potential is

$$(V_p\hat{\alpha})(x, p) = \int_{\partial S} \hat{D}(x - y, p)\hat{\alpha}(y, p)\, ds_y.$$

We remark that, since $\hat{D}(x, p) = \hat{D}^{\mathrm{T}}(-x, p)$, the above equality can also be written in the form

$$(V_p\hat{\alpha})(x, p) = \int_{\partial S} (\hat{\alpha}(y, p), \hat{D}^{(j)}(y - x, p)) e_j\, ds_y,$$

where e_j are the coordinate unit vectors. The main properties of $V_p\hat{\alpha}$ for $\hat{\alpha} \in C^2(\partial S)$ are listed in §1.3. On the basis of these properties, we can define a boundary operator $V_{p,0}$ by setting

$$V_{p,0}\hat{\alpha} = (V_p\hat{\alpha})_0 = \gamma^{\pm}\pi^{\pm} V_p\hat{\alpha}, \quad p \in \mathbb{C}_0.$$

4.1 Lemma. *For any $p \in \mathbb{C}_0$, the operator $V_{p,0}$, extended by continuity from $C^2(\partial S)$ to $H_{-1/2,p}(\partial S)$, is a homeomorphism from $H_{-1/2,p}(\partial S)$ to $H_{1/2,p}(\partial S)$, and for any $\hat{\alpha} \in H_{-1/2,p}(\partial S)$, $p \in \bar{\mathbb{C}}_\kappa$,*

$$\|V_{p,0}\hat{\alpha}\|_{1/2,p;\partial S} \leq c|p|\|\hat{\alpha}\|_{-1/2,p;\partial S}, \tag{4.1}$$

$$\|\hat{\alpha}\|_{-1/2,p;\partial S} \leq c|p|\|V_{p,0}\hat{\alpha}\|_{1/2,p;\partial S}, \tag{4.2}$$

$$\|\pi^+ V_p\hat{\alpha}\|_{1,p;S^+} + \|\pi^- V_p\hat{\alpha}\|_{1,p;S^-} \leq c|p|\|\hat{\alpha}\|_{-1/2,p;\partial S}. \tag{4.3}$$

Proof. Let $\hat{\alpha} \in C^2(\partial S)$. From the properties of $V_p\hat{\alpha}$ it follows that $V_{p,0}\hat{\alpha} \in H_{1/2,p}(\partial S)$ and $\pi^\pm V_p\hat{\alpha} \in H_{1,p}(S^\pm)$; consequently,

$$\|V_{p,0}\hat{\alpha}\|_{1/2,p;\partial S}^2 \leq c\|\pi^+ V_p\hat{\alpha}\|_{1,p;S^+}^2$$

$$\leq c(\|\pi^+ V_p\hat{\alpha}\|_{1,p;S^+}^2 + \|\pi^- V_p\hat{\alpha}\|_{1,p;S^-}^2). \tag{4.4}$$

Since $\pi^\pm V_p\hat{\alpha}$ are the solutions of (3.1) with $f = V_{p,0}\hat{\alpha}$ in $S^\pm$, respectively, (3.9) implies that for $p \in \bar{\mathbb{C}}_\kappa$,

$$\|\pi^+ V_p\hat{\alpha}\|_{1,p;S^+}^2 + \|\pi^- V_p\hat{\alpha}\|_{1,p;S^-}^2$$

$$\leq c \operatorname{Re}\left\{\bar{p}\left((\mathcal{T}_p^+ - \mathcal{T}_p^-)V_{p,0}\hat{\alpha}, V_{p,0}\hat{\alpha}\right)_{0;\partial S}\right\}.$$

We remark that the jump formula for the single-layer potential $V_p\hat{\alpha}$ takes the form

$$(\mathcal{T}_p^+ - \mathcal{T}_p^-)V_{p,0}\hat{\alpha} = \hat{\alpha}; \tag{4.5}$$

hence,

$$\|\pi^+ V_p\hat{\alpha}\|_{1,p;S^+}^2 + \|\pi^- V_p\hat{\alpha}\|_{1,p;S^-}^2 \leq c|p||(\hat{\alpha}, V_{p,0}\hat{\alpha})_{0;\partial S}|. \tag{4.6}$$

By (4.4) and (4.6),

$$\|V_{p,0}\hat{\alpha}\|_{1/2,p;\partial S}^2 \leq c|p|\|V_{p,0}\hat{\alpha}\|_{1/2,p;\partial S}\|\hat{\alpha}\|_{-1/2,p;\partial S},$$

so

$$\|V_{p,0}\hat{\alpha}\|_{1/2,p;\partial S} \leq c|p|\|\hat{\alpha}\|_{-1/2,p;\partial S} \quad \forall \hat{\alpha} \in C^2(\partial S). \tag{4.7}$$

Furthermore, from (4.7) it follows that

$$\|\pi^+ V_p\hat{\alpha}\|_{1,p;S^+} + \|\pi^- V_p\hat{\alpha}\|_{1,p;S^-} \leq c|p|\|\hat{\alpha}\|_{-1/2,p;\partial S}.$$

Inequality (4.7) shows that $V_{p,0}$ can be extended by continuity to $H_{-1/2,p}(\partial S)$, that the extended operator is continuous from the space $H_{-1/2,p}(\partial S)$ to $H_{1/2,p}(\partial S)$, that (4.1) holds because the operators $\mathcal{T}_p^\pm$ are continuous from $H_{1/2,p}(\partial S)$ to $H_{-1/2,p}(\partial S)$, and that (4.5) remains valid for all densities $\hat{\alpha} \in H_{-1/2,p}(\partial S)$. From the above estimates it follows that (4.3) is also valid for any $\hat{\alpha} \in H_{-1/2,p}(\partial S)$. The equality

$$\gamma^+ \pi^+ V_p\hat{\alpha} = \gamma^- \pi^- V_p\hat{\alpha} = V_{p,0}\hat{\alpha},$$

which holds for smooth densities, also holds for any $\hat{\alpha} \in H_{-1/2,p}(\partial S)$. By (4.5), $V_{p,0}$ is injective from $H_{-1/2,p}(\partial S)$ to $H_{1/2,p}(\partial S)$ and

$$V_{p,0}^{-1} = \mathcal{T}_p^+ - \mathcal{T}_p^-.$$

This formula and (3.3) imply that $V_{p,0}^{-1}$ is continuous from $H_{1/2,p}(\partial S)$ to $H_{-1/2,p}(\partial S)$, and that (4.2) holds.

To complete the proof, we need to check that the range of $V_{p,0}$ is dense in $H_{1/2,p}(\partial S)$. If we assume the opposite, then there is a nonzero $\varphi \in H_{-1/2,p}(\partial S)$ such that

$$(V_{p,0}\hat{\alpha}, \varphi)_{0;\partial S} = 0 \quad \forall \hat{\alpha} \in H_{-1/2,p}(\partial S).$$

Taking $\hat{\alpha} = \varphi$, from (4.6) we see that $\pi^{\pm} V_p \varphi = 0$; therefore,

$$V_{p,0}\varphi = \gamma^+ \pi^+ V_p \varphi = \gamma^- \pi^- V_p \varphi = 0,$$

so $\varphi = 0$. This contradiction proves the assertion. $\qquad \square$

Before returning to the spaces of originals, we need an auxiliary result. We consider the problem of finding $u \in H_{1,p}(\mathbb{R}^2)$ such that

$$p^2(B^{1/2}u, B^{1/2}v)_0 + a(u, v) = (q, v)_0 + (g, \gamma v)_{0;\partial S} \quad \forall v \in H_{1,p}(\mathbb{R}^2), \quad (4.8)$$

where

$$a(u, v) = 2 \int_{\mathbb{R}^2} E(u, v) \, dx,$$

γv is the trace of v on ∂S, and $q \in H_{-1,p}(\mathbb{R}^2)$ and $g \in H_{-1/2,p}(\partial S)$ are prescribed.

4.2 Lemma. *For any given* $q \in H_{-1,p}(\mathbb{R}^2)$ *and* $g \in H_{-1/2,p}(\partial S)$, $p \in \mathbb{C}_0$, *problem (4.8) has a unique solution* $u \in H_{1,p}(\mathbb{R}^2)$, *and for any* $p \in \bar{\mathbb{C}}_\kappa$, $\kappa > 0$,

$$\|u\|_{1,p} \leq c|p|(\|q\|_{-1,p} + \|g\|_{-1/2,p;\partial S}). \quad (4.9)$$

Proof. First, we remark that $(g, \gamma v)_{0;\partial S}$ defines a bounded antilinear (conjugate linear) functional on $H_{1,p}(\mathbb{R}^2)$ since, by the trace theorem,

$$|(g, \gamma v)_{0;\partial S}| \leq \|g\|_{-1/2,p;\partial S} \|\gamma v\|_{1/2,p;\partial S}$$
$$\leq c\|g\|_{-1/2,p;\partial S} \|v\|_{1,p},$$

so there is $\tilde{q} \in H_{-1,p}(\mathbb{R}^2)$ such that $\|\tilde{q}\|_{-1,p} \leq c\|g\|_{-1/2,p;\partial S}$ and

$$(g, \gamma v)_{0;\partial S} = (\tilde{q}, v)_0 \quad \forall v \in H_{1,p}(\mathbb{R}^2).$$

Writing $Q = q + \tilde{q} \in H_{-1,p}(\mathbb{R}^2)$, we bring (4.8) to the form

$$p^2(B^{1/2}u, B^{1/2}v)_0 + a(u,v) = (Q,v)_0 \quad \forall v \in H_{1,p}(\mathbb{R}^2). \tag{4.10}$$

We now fix $\kappa > 0$ and repeat the corresponding part of the proof of Theorem 2.2. Thus, we introduce the form

$$a_\kappa(u,v) = \tfrac{1}{2}\kappa^2(B^{1/2}u, B^{1/2}v)_0 + a(u,v),$$

which, clearly, is continuous, symmetric, and coercive on $[H_{1,p}(\mathbb{R}^2)]^2$. This form defines a self-adjoint nonnegative operator A_κ by

$$(A_\kappa u, v)_0 = a_\kappa(u,v) \quad \forall u, v \in H_{1,p}(\mathbb{R}^2).$$

Then we rewrite (4.10) in the form

$$A_\kappa u + (p^2 - \tfrac{1}{2}\kappa^2)Bu = Q$$

and prove the unique solvability of this equation in $H_{1,p}(\mathbb{R}^2)$, just as we did in Theorem 2.2.

Taking $v = u$ in (4.10) and separating the real and imaginary parts, we obtain

$$|p|^2 \|B^{1/2}u\|_0^2 + a(u,u) = \sigma^{-1}\operatorname{Re}\{\bar{p}(Q,u)_0\};$$

therefore,

$$\|u\|_{1,p} \le c|p|\|Q\|_{-1,p} \le c|p|(\|q\|_{-1,p} + \|g\|_{-1/2,p;\partial S}),$$

which proves (4.9). □

We now define a pair of operators $\hat{V}_0$ and $\hat{V}_0^{-1}$ on the elements $\tilde{\alpha}(p) = \hat{\alpha}(\cdot, p)$ and $F(p) = \hat{f}(\cdot, p)$, $p \in \mathbb{C}_\kappa$, of $H_{-1/2,k,\kappa}(\partial S)$ and $H_{1/2,k,\kappa}(\partial S)$, respectively, by

$$(\hat{V}_0\tilde{\alpha})(p) = (V_{p,0}\hat{\alpha})(\cdot, p),$$
$$(\hat{V}_0^{-1}F)(p) = (V_{p,0}^{-1}\hat{f})(\cdot, p).$$

We also define the single-layer potential $\hat{V}\tilde{\alpha}$ by

$$(\hat{V}\tilde{\alpha})(x,p) = (V_p\hat{\alpha})(x,p), \quad x \in \mathbb{R}^2,\ p \in \mathbb{C}_\kappa.$$

Returning to the spaces of originals, we define operators V_0 and V_0^{-1} and the single-layer potential $V\alpha$ by setting

$$V_0\alpha = \mathcal{L}^{-1}\hat{V}_0\mathcal{L}\alpha,$$
$$V_0^{-1}f = \mathcal{L}^{-1}\hat{V}_0^{-1}\mathcal{L}f,$$
$$(V\alpha)(X) = (\mathcal{L}^{-1}\hat{V}\tilde{\alpha})(X), \quad X \in \mathbb{R}^2 \times \mathbb{R}.$$

4.3 Theorem. *For any $\kappa > 0$ and $k \in \mathbb{R}$, the operator V_0 is continuous and injective from $H_{-1/2,k,\kappa}(\Gamma)$ to $H_{1/2,k-1,\kappa}(\Gamma)$, and its range is dense in $H_{1/2,k-1,\kappa}(\Gamma)$. The inverse V_0^{-1}, extended by continuity from the range of V_0, is continuous and injective from $H_{1/2,k,\kappa}(\Gamma)$ to $H_{-1/2,k-1,\kappa}(\Gamma)$ for any $k \in \mathbb{R}$, and its range is dense in $H_{-1/2,k-1,\kappa}(\Gamma)$. In addition, for any $\alpha \in H_{-1/2,k,\kappa}(\Gamma)$,*

$$\|\pi^+ V\alpha\|_{1,k-1,\kappa;G^+} + \|\pi^- V\alpha\|_{1,k-1,\kappa;G^-} \leq c\|\alpha\|_{-1/2,k,\kappa;\Gamma}. \tag{4.11}$$

Proof. Let $\tilde{\alpha}(p) \in H_{-1/2,k,\kappa}(\partial S)$, $p \in \mathbb{C}_\kappa$. Since $V_p\hat{\alpha}$ is the solution of (4.8) with

$$q = 0, \quad g = (\mathcal{T}_p^+ - \mathcal{T}_p^-)V_{p,0}\hat{\alpha} = \hat{\alpha},$$

from (4.9) it follows that

$$\|\pi^+ V_p\hat{\alpha}\|_{1,p;S^+} + \|\pi^- V_p\hat{\alpha}\|_{1,p;S^-} \leq c|p|\|\hat{\alpha}\|_{-1/2,p;\partial S}.$$

We claim that $\pi^\pm V_p\hat{\alpha}(\cdot,p)$ are holomorphic mappings from $\mathbb{C}_\kappa$ to $H_1(S^\pm)$, respectively. To show this, we choose any $p_0 \in \mathbb{C}_\kappa$, consider a circle $K_R(p_0)$ with the center at p_0 and radius R such that $\bar{K}_R(p_0) \subset \mathbb{C}_\kappa$, and rewrite (4.8) as

$$p_0^2(B^{1/2}V_p\hat{\alpha}, B^{1/2}v)_0 + a(V_p\hat{\alpha}, v)$$
$$= -(p^2 - p_0^2)(B^{1/2}V_p\hat{\alpha}, B^{1/2}v)_0 + (\hat{\alpha}, \gamma v)_{0;\partial S} \quad \forall v \in H_{1,p}(\mathbb{R}^2).$$

Taking (4.9) into account and repeating the proof of Theorem 3.2, we conclude, successively, that $V_p\hat{\alpha}(\cdot,p)$ is a bounded, continuous, and holomorphic mapping from $\mathbb{C}_\kappa$ to $H_1(\mathbb{R}^2)$, and that, consequently, $\pi^\pm V_p\hat{\alpha}(\cdot,p)$ are holomorphic mappings from $\mathbb{C}_\kappa$ to $H_1(S^\pm)$; therefore,

$$V_{p,0}\hat{\alpha} = \gamma^+ \pi^+ V_p\hat{\alpha} = \gamma^- \pi^- V_p\hat{\alpha}$$

is holomorphic from $\mathbb{C}_\kappa$ to $H_{1/2}(\partial S)$. Since

$$\|V_0\alpha\|_{1/2,k-1,\kappa;\Gamma}^2 = \sup_{\sigma > \kappa} \int_{-\infty}^{\infty} (1 + |p|^2)^{k-1}\|V_{p,0}\hat{\alpha}\|_{1/2,p;\partial S}^2 \, d\tau$$

$$\leq c \sup_{\sigma > \kappa} \int_{-\infty}^{\infty} (1 + |p|^2)^k\|\hat{\alpha}\|_{-1/2,p;\partial S}^2 \, d\tau$$

$$= \|\alpha\|_{-1/2,k,\kappa;\Gamma}^2, \quad p = \sigma + i\tau,$$

the mapping $V_0 : H_{-1/2,k,\kappa}(\Gamma) \to H_{1/2,k-1,\kappa}(\Gamma)$ is continuous for any $k \in \mathbb{R}$. If $V_0\hat{\alpha} = 0$, then $(\hat{V}_0\tilde{\alpha})(p) = 0$, $\tilde{\alpha}(p) = \hat{\alpha}(\cdot, p)$, which means that $(V_{p,0}\hat{\alpha})(\cdot, p) = 0$ for every $p \in \mathbb{C}_\kappa$; hence, $\hat{\alpha} = 0$.

The continuity of V_0^{-1} follows from the equality

$$V_0^{-1} = \mathcal{T}^+ - \mathcal{T}^-$$

and Theorem 3.2. Finally, the statement that the ranges of V_0 and V_0^{-1} are dense in the corresponding spaces follows from the fact that $H_{\pm 1/2,k+1,\kappa}(\Gamma)$ are dense in $H_{\pm 1/2,k-1,\kappa}(\Gamma)$, respectively. Inequality (4.11) is obtained from (4.3), and the assertion is proved.   □

We now go over to the properties of the (retarded) double-layer potential $W\beta$ with density β defined on Γ. This potential was introduced in §1.3 for smooth densities as

$$(W\beta)(X) = \int_0^\infty \int_{\partial S} P(x - y, t - \tau)\beta(y, \tau)\, ds_y\, d\tau$$

$$= \int_0^\infty \int_{\partial S} (\beta(y, \tau), (T_y D^{(j)})(y - x, t - \tau))e_j\, ds_y\, d\tau,$$

and its Laplace transform is

$$(W_p\hat{\beta})(x, p) = \int_{\partial S} \hat{P}(x - y, p)\hat{\beta}(y, p)\, ds_y$$

$$= \int_{\partial S} (\hat{\beta}(y, p), (T_y \hat{D}^{(j)})(y - x, p))e_j\, ds_y.$$

4.4 Lemma. *For any $p \in \mathbb{C}_0$ and $\hat{\beta} \in C^2(\partial S)$, the double-layer potential $W_p\hat{\beta}$ can be written in the form*

$$(W_p\hat{\beta})(x, p) = \begin{cases} (V_p\mathcal{T}_p^- \hat{\beta})(x, p), & x \in S^+, \\ (V_p\mathcal{T}_p^+ \hat{\beta})(x, p), & x \in S^-. \end{cases} \tag{4.12}$$

Proof. Let $x \in S^+$ be fixed, and consider $\hat{D}^{(j)}(y - x, p)$ as a function of $y \in \bar{S}^-$. Also, let $f, \varphi \in H_{1/2}(\partial S)$, and let u and v be the solutions of the exterior problem (3.1) with boundary data f and φ, respectively. By the definition of $\mathcal{T}^-$,

$$-(\mathcal{T}_p^- f, \bar{\varphi})_{0;\partial S} = p^2(B^{1/2}u, B^{1/2}\bar{v})_{0;S^-} + a_-(u, \bar{v}),$$

$$-(\mathcal{T}_p^- \varphi, \bar{f})_{0;\partial S} = p^2(B^{1/2}v, B^{1/2}\bar{u})_{0;S^-} + a_-(v, \bar{u}).$$

Since the right-hand sides above are equal, we deduce that

$$(T_p^- f, \bar{\varphi})_{0;\partial S} = (T_p^- \varphi, \bar{f})_{0;\partial S},$$

or

$$\int_{\partial S} \left((T_p^- f)(y), \varphi(y) \right) ds_y = \int_{\partial S} \left(f(y), (T_p^- \varphi)(y) \right) ds_y.$$

Taking $f(y) = \hat{\beta}(y, p)$ and $\varphi(y) = \hat{D}^{(j)}(y - x, p)$ in this equality, we find that

$$\int_{\partial S} \left((T_p^- \hat{\beta})(y, p), \hat{D}^{(j)}(y - x, p) \right) ds_y$$
$$= \int_{\partial S} \left(\hat{\beta}(y, p), (T_p^- \hat{D}^{(j)})(y - x, p) \right) ds_y.$$

Multiplying both sides above by e_j and summing over j from 1 to 3, we obtain the first equality (4.12). The second one is derived analogously.   □

From the properties of the double-layer potential with a smooth density it follows that $\pi^\pm W_p \hat{\beta} \in H_{1,p}(S^\pm)$. Consequently, we can define the operators $W_p^\pm$ of the limiting values on ∂S of $W_p \hat{\beta}$ with $\hat{\beta} \in C^2(\partial S)$ by

$$W_p^\pm \hat{\beta} = \gamma^\pm \pi^\pm W_p \hat{\beta}.$$

4.5 Lemma. *For any $p \in \mathbb{C}_0$, the double-layer potential $W_p \hat{\beta}$ and the operators $W_p^\pm$ of its limiting values can be extended by continuity from $C^2(\partial S)$ to $H_{1/2,p}(\partial S)$. The extended operators $W_p^\pm$ are homeomorphisms from the space $H_{1/2,p}(\partial S)$ to $H_{1/2,p}(\partial S)$, and the extended functions $\pi^\pm W_p \hat{\beta}$ are continuous from $H_{1/2,p}(\partial S)$ to $H_{1,p}(S^\pm)$, respectively. In addition, for any $\hat{\beta} \in H_{1/2,p}(\partial S)$, $p \in \mathbb{C}_\kappa$, $\kappa > 0$,*

$$\|\pi^+ W_p \hat{\beta}\|_{1,p;S^+} + \|\pi^- W_p \hat{\beta}\|_{1,p;S^-} \leq c|p|^2 \|\hat{\beta}\|_{1/2,p;\partial S}, \tag{4.13}$$

$$\|W_p^\pm \hat{\beta}\|_{1/2,p;\partial S} \leq c|p|^2 \|\hat{\beta}\|_{1/2,p;\partial S}, \tag{4.14}$$

$$\|\hat{\beta}\|_{1/2,p;\partial S} \leq c|p|^2 \|W_p^\pm \hat{\beta}\|_{1/2,p;\partial S}. \tag{4.15}$$

This assertion and inequalities (4.13)–(4.15) follow from Lemmas 3.1 and 4.1.

We now define pairs of operators $\hat{W}^\pm$ and $(\hat{W}^\pm)^{-1}$ on the elements $\tilde{\beta}(p) = \hat{\beta}(\cdot, p)$ and $F(p) = \hat{f}(\cdot, p)$, $p \in \mathbb{C}_\kappa$, of $H_{1/2,k,\kappa}(\partial S)$ by

$$(\hat{W}^\pm \tilde{\beta})(p) = (W_p^\pm \hat{\beta})(\cdot, p),$$
$$((\hat{W}^\pm)^{-1} F)(p) = ((W_p^\pm)^{-1} \hat{f})(\cdot, p),$$

and the double-layer potential $\hat{W}\tilde{\beta}$ by

$$(\hat{W}\tilde{\beta})(x,p) = (W_p\hat{\beta})(x,p), \quad x \in S^\pm, \; p \in \mathbb{C}_\kappa.$$

Returning to the spaces of originals, we define operators $W^\pm$ and $(W^\pm)^{-1}$ and the double-layer potential $W\beta$ by setting

$$W^\pm\beta = \mathcal{L}^{-1}\hat{W}^\pm\mathcal{L}\beta,$$
$$(W^\pm)^{-1}f = \mathcal{L}^{-1}(\hat{W}^\pm)^{-1}\mathcal{L}f,$$
$$(W\beta)(X) = (\mathcal{L}^{-1}\hat{W}\tilde{\beta})(X), \quad X \in G^+ \cup G^-.$$

4.6 Theorem. *For any $\kappa > 0$ and $k \in \mathbb{R}$, the operators $W^\pm$ are continuous and injective from $H_{1/2,k,\kappa}(\Gamma)$ to $H_{1/2,k-2,\kappa}(\Gamma)$. Their inverses $(W^\pm)^{-1}$, extended by continuity from the ranges of $W^\pm$, respectively, to $H_{1/2,k,\kappa}(\Gamma)$, are continuous and injective operators from $H_{1/2,k,\kappa}(\Gamma)$ to $H_{1/2,k-2,\kappa}(\Gamma)$ for any $k \in \mathbb{R}$, and their ranges are dense in $H_{1/2,k-2,\kappa}(\Gamma)$. In addition, for any $\beta \in H_{1/2,k,\kappa}(\Gamma)$,*

$$\|\pi^+W\beta\|_{1,k-2,\kappa;G^+} + \|\pi^-W\beta\|_{1,k-2,\kappa;G^-} \le c\|\beta\|_{1/2,k,\kappa;\Gamma}.$$

Proof. The assertion follows from the equalities

$$(W\beta)(X) = \begin{cases} (V\mathcal{T}^-\beta)(X), & X \in G^+, \\ (V\mathcal{T}^+\beta)(X), & X \in G^-, \end{cases}$$
$$W^\pm\beta = V_0\mathcal{T}^\mp\beta,$$

and Theorems 3.2 and 4.3.   □

4.7 Lemma. *For any $\beta \in H_{1/2,k,\kappa}(\Gamma)$, $\kappa > 0$, $k \in \mathbb{R}$,*

$$W^+\beta - W^-\beta = -\beta, \tag{4.16}$$
$$\mathcal{T}^+W^+\beta = \mathcal{T}^-W^-\beta. \tag{4.17}$$

Proof. Let $p \in \mathbb{C}_\kappa$. By (4.5),

$$W_p^+\hat{\beta} - W_p^-\hat{\beta} = V_{p,0}(\mathcal{T}_p^- - \mathcal{T}_p^+)\hat{\beta} = -\hat{\beta} \tag{4.18}$$

and

$$\begin{aligned} \mathcal{T}_p^+W_p^+\hat{\beta} &= \mathcal{T}_p^+V_{p,0}\mathcal{T}_p^-\hat{\beta} \\ &= \mathcal{T}_p^-\hat{\beta} + \mathcal{T}_p^-V_{p,0}\mathcal{T}_p^-\hat{\beta} \\ &= \mathcal{T}_p^-V_{p,0}\mathcal{T}_p^+\hat{\beta} = \mathcal{T}_p^-W_p^-\hat{\beta}. \end{aligned} \tag{4.19}$$

It is now easy to see that (4.16) and (4.17) follow from (4.18) and (4.19), respectively.   □

We write $N_p = T_p^+ W_p^+ = T_p^- W_p^-$ and use the above procedure to define the operators $\hat{N}$, $\hat{N}^{-1}$ and N, N^{-1}.

4.8 Theorem.   *For any $\kappa > 0$ and $k \in \mathbb{R}$, the operator N is continuous and injective from $H_{1/2,k,\kappa}(\Gamma)$ to $H_{-1/2,k-3,\kappa}(\Gamma)$, and its range is dense in $H_{-1/2,k-3,\kappa}(\Gamma)$. Its inverse N^{-1}, extended by continuity from the range of N to $H_{-1/2,k,\kappa}(\Gamma)$, is continuous and injective from the space $H_{-1/2,k,\kappa}(\Gamma)$ to $H_{1/2,k-1,\kappa}(\Gamma)$ for any $k \in \mathbb{R}$, and its range is dense in $H_{1/2,k-1,\kappa}(\Gamma)$.*

Proof. The assertion concerning N follows from Theorems 3.2 and 4.6. Alternatively, it can be established from the estimate

$$\|N_p\hat{\beta}\|_{-1/2,p;\partial S} \le c|p|^3\|\hat{\beta}\|_{1/2,p;\partial S},$$

which is obtained from (3.3) and (4.14).

Let $\hat{\beta} \in H_{1/2,p}(\partial S)$ and $p \in \bar{\mathbb{C}}_\kappa$. By (4.18), (4.19), and the trace theorem,

$$\|\hat{\beta}\|^2_{1/2,p;\partial S} = \|W_p^- \hat{\beta} - W_p^+ \hat{\beta}\|^2_{1/2,p;\partial S}$$

$$\le c(\|W_p^+ \hat{\beta}\|^2_{1/2,p;\partial S} + \|W_p^- \hat{\beta}\|^2_{1/2,p;\partial S})$$

$$\le c(\|\pi^+ W_p\hat{\beta}\|^2_{1,p;S^+} + \|\pi^- W_p\hat{\beta}\|^2_{1,p;S^-})$$

$$\le c|p||(W_p^+ \hat{\beta} - W_p^- \hat{\beta}, N_p\hat{\beta})_{0;\partial S}|$$

$$\le c|p|\|N_p\hat{\beta}\|_{-1/2,p;\partial S}\|\hat{\beta}\|_{1/2,p;\partial S};$$

therefore,

$$\|\hat{\beta}\|_{1/2,p;\partial S} \le c|p|\|N_p\hat{\beta}\|_{-1/2,p;\partial S}.$$

This inequality and Theorems 3.2 and 4.6 complete the proof.   □

4.2 Nonstationary Boundary Integral Equations

Problems (DD$^\pm$) consist in solving

$$B(\partial_t^2 u)(X) + (Au)(X) = 0, \quad X \in G^\pm,$$

$$u(x,0+) = (\partial_t u)(x,0+) = 0, \quad x \in S^\pm,$$

$$u^\pm(X) = f(X), \quad X \in \Gamma,$$

where f is prescribed on Γ. We seek their solutions in the form

$$u(X) = (V\alpha)(X), \quad X \in G^+ \text{ or } X \in G^-, \tag{4.20}$$

where α is an unknown density defined on $\partial S \times \mathbb{R}$, which is zero for $t < 0$. Alternatively, we can write

$$u = \pi^\pm V \alpha.$$

This representation yields the system of nonstationary boundary integral equations

$$V_0 \alpha = f, \tag{4.21}$$

where V_0 is the boundary operator defined by the single-layer potential (see §4.1). On the other hand, the representation

$$u(X) = (W\beta)(X), \quad X \in G^+ \text{ or } x \in G^-, \tag{4.22}$$

or, which is the same,

$$u = \pi^\pm W\beta,$$

of the solutions of these problems yields the boundary systems

$$W^\pm \beta = f, \tag{4.23}$$

where $W^\pm$ are the operators of the limiting values on Γ of the double-layer potential and β is an unknown density defined on $\partial S \times \mathbb{R}$ and vanishing for $t < 0$.

4.9 Theorem. *For any $f \in H_{1/2,k,\kappa}(\Gamma)$, $\kappa > 0$, $k \in \mathbb{R}$, systems (4.21) and (4.23) have unique solutions $\alpha \in H_{-1/2,k-1,\kappa}(\Gamma)$ and $\beta \in H_{1/2,k-2,\kappa}(\Gamma)$, in which case the functions u defined by (4.20) or (4.22) belong to $H_{1,k-1,\kappa}(G^\pm)$. If $k \geq 1$, then these functions are the weak solutions of problems $(DD^\pm)$, respectively.*

Proof. The assertion concerning the solvability of (4.21) and (4.23) follows from Theorems 4.3 and 4.6.

Now let α be the solution of (4.21) with $f \in H_{1/2,k,\kappa}(\Gamma)$. Then $\pi^\pm V_p \hat{\alpha} \in H_{1,p}(S^\pm)$ for any $p \in \mathbb{C}_\kappa$ and are infinitely differentiable with respect to $x \in S^\pm$. A straightforward calculation shows that these functions satisfy

$$p^2\left(B^{1/2}(\pi^\pm V_p\hat{\alpha}), B^{1/2}v\right) + a_\pm(u, v) = 0 \quad \forall v \in C_0^\infty(S^\pm),$$
$$\gamma^\pm \pi^\pm V_p\hat{\alpha} = V_{p,0}\hat{\alpha} = \hat{f}. \tag{4.24}$$

It is clear that (4.24) also hold for all $v \in \mathring{H}_{1,p}(S^\pm)$. Furthermore, by (4.6),

$$\|\pi^+ V_p\hat{\alpha}\|^2_{1,p;S^+} + \|\pi^- V_p\hat{\alpha}\|^2_{1,p;S^-} \leq c|p|\left|(\hat{f}, V_{p,0}^{-1}\hat{f})_{0;\partial S}\right|,$$

so (4.2) implies that

$$\|\pi^+ V_p\hat{\alpha}\|_{1,p;S^+} + \|\pi^- V_p\hat{\alpha}\|_{1,p;S^-} \leq c|p|\|\hat{f}\|_{1/2,p;\partial S}.$$

Taking this inequality into account and repeating the proof of Theorem 2.3, we arrive at the required assertion. The case of (4.23) is treated similarly. □

We now represent the solutions of problems (DN$^\pm$) in the form (4.20) or (4.22). These representations yield, respectively, the systems of nonstationary boundary integral equations

$$T^\pm V_0 \alpha = g, \qquad (4.25)$$

$$N\beta = g. \qquad (4.26)$$

4.10 Theorem. *For any $g \in H_{-1/2,k,\kappa}(\Gamma)$, $\kappa > 0$, $k \in \mathbb{R}$, systems (4.25) and (4.26) have unique solutions $\alpha \in H_{-1/2,k-2,\kappa}(\Gamma)$ and $\beta \in H_{1/2,k-1,\kappa}(\Gamma)$, in which case the functions u defined by (4.20) or (4.22) belong to the space $H_{1,k-1,\kappa}(G^\pm)$. If $k \geq 1$, then these functions are the weak solutions of problems (DN$^\pm$), respectively.*

The proof of this assertion is a repeat of that of Theorem 4.9, use being made of Theorems 3.2, 4.3, and 4.8.

4.3 The Direct Method

We begin by establishing the dynamic analog of the third Green's formula (or Somigliana representation formula, as is known in elasticity theory). Let $u \in H_{1,k,\kappa}(G^+)$, $k \in \mathbb{R}$, $\kappa > 0$, be the weak solution of problem (DD$^+$)

$$B(\partial^2 u)(X) + (Au)(X) = 0, \quad X \in G^+,$$
$$u(x,0+) = (\partial_t u)(x,0+) = 0, \quad x \in S^+, \qquad (4.27)$$
$$(\gamma^+ u)(X) = f(X), \quad X \in \Gamma,$$

where $f \in H_{1/2,k,\kappa}(\Gamma)$. Then for any $p \in \mathbb{C}_\kappa$, $\hat{u}(x,p) = \mathcal{L}u(x,t)$ is the weak solution of the problem

$$Bp^2 \hat{u}(x,p) + (A\hat{u})(x,p) = 0, \quad x \in S^+,$$
$$(\gamma^+ \hat{u})(x,p) = \hat{f}(x,p), \quad x \in \partial S, \qquad (4.28)$$

where $\hat{f} \in H_{1/2,p}(\partial S)$. We now consider the function

$$w = \pi^+ V_p T_p^+ \hat{f} - \pi^+ W_p \hat{f},$$

which belongs to $H_{1,p}(S^+)$. Given that

$$\gamma^+ w = V_{p,0} T_p^+ \hat{f} - W_p^+ \hat{f} = V_{p,0}(T_p^+ - T_p^-)\hat{f} = \hat{f},$$

w is also a weak solution of (4.28). Since (4.28) has a unique solution in $H_{1,p}(S^+)$, we conclude that

$$\hat{u}(x,p) = (\pi^+ V_p T_p^+ \hat{f})(x,p) - (\pi^+ W_p \hat{f})(x,p), \quad x \in S^+.$$

Returning to the spaces of originals, we obtain

$$u(X) = (\pi^+ V T^+ f)(X) - (\pi^+ W f)(X), \quad X \in G^+. \tag{4.29}$$

From this it follows that $f = V_0 T^+ f - W^+ f$. Setting $\alpha = T^+ f$ and taking into account that $f + W^+ f = W^- f$, we arrive at

$$V_0 \alpha = W^- f. \tag{4.30}$$

This nonstationary boundary integral equation for the density α has a direct physical meaning, since α is the boundary moment-force field. Solving (4.30), we obtain the solution of (4.27) as

$$u(X) = (\pi^+ V \alpha)(X) - (\pi^+ W f)(X), \quad X \in G^+.$$

In the case of the exterior problem (DD$^-$), it is not difficult to show that representation (4.29) takes the form

$$u(X) = -(\pi^- V T^- f)(X) + (\pi^- W f)(X), \quad X \in G^-, \tag{4.31}$$

which leads to the system of boundary integral equations

$$V_0 \alpha = W^+ f \tag{4.32}$$

for the density $\alpha = T^- f$.

4.11 Theorem. *For any $f \in H_{1/2,k,\kappa}(\Gamma)$, $\kappa > 0$, $k \in \mathbb{R}$, systems (4.30) and (4.32) have unique solutions $\alpha \in H_{-1/2,k-1,\kappa}(\Gamma)$, in which case*

$$u = \pm(\pi^\pm V \alpha - \pi^- W f) \in H_{1,k-1,\kappa}(G^\pm).$$

If $k \geq 1$, then these functions u are the weak solutions of problems (DD$^\pm$), respectively.

The assertion is proved by using the representations $\alpha = T^\pm f$ and the properties of the boundary operators defined in §4.1.

Going over to (DN$^\pm$), we remark that, since here the boundary moment-force vector $T^\pm \gamma^\pm u = g$ is known, we write the representations (4.29) and (4.31) as

$$u(X) = \pm \big[(\pi^\pm V g)(X) - (\pi^\pm W \beta)(X) \big], \quad X \in G^\pm, \tag{4.33}$$

where $\beta = \gamma^\pm u$ is the unknown displacement field on the boundary. Since $\gamma^\pm u = \pm(V_0 g - W^\pm \beta)$, we arrive at the systems of boundary integral equations

$$N\beta = \mathcal{T}^\mp V_0 g. \tag{4.34}$$

4.12 Theorem. *For any $g \in H_{-1/2,k,\kappa}(\Gamma)$, $\kappa > 0$, $k \in \mathbb{R}$, systems (4.34) have unique solutions $\beta \in H_{1/2,k-1,\kappa}(\Gamma)$, in which case the functions u defined by (4.33) belong to $H_{1,k-1,\kappa}(G^\pm)$. If $k \geq 1$, then these functions are the weak solutions of problems* $(\mathrm{DN}^\pm)$, *respectively.*

The assertion follows from the representations $\beta = (\mathcal{T}^\pm)^{-1} g$ and the properties of the boundary operators discussed above.

5

Transmission Problems and Multiply Connected Plates

5.1 Infinite Plate with a Finite Inclusion

Suppose that S^+ and S^- are the middle plane domains of plates with different Lamé constants λ_+, μ_+ and λ_-, μ_-, different densities ρ_+, ρ_-, and different thickness parameters h_+, h_-. Let B_+ and B_- be the corresponding diagonal matrices B. Also, let $A_\pm$, $a_\pm$, and $T_\pm$ be, respectively, the matrix differential operators, internal energy bilinear forms, and boundary moment-force operators associated with $S^\pm$.

If $u \in H_{1,k,\kappa}(\mathbb{R}^2 \times (0,\infty))$, then we make the notation

$$u_\pm = \pi^\pm u \in H_{1,k,\kappa}(G^\pm), \quad u = \{u_+, u_-\}.$$

Similarly, for the Laplace transform $\hat{u} \in H_{1,p}(\mathbb{R}^2)$ of u we write

$$\hat{u} = \{\hat{u}_+, \hat{u}_-\}, \quad \hat{u}_\pm = \pi^\pm \hat{u} \in H_{1,p}(S^\pm).$$

In §1.1 we stated that the classical transmission (contact) problem (DT) consists in finding

$$u_+ \in C^2(G^+) \cap C^1(\bar{G}^+), \quad u_- \in C^2(G^-) \cap C^1(\bar{G}^-)$$

satisfying

$$
\begin{aligned}
B_+(\partial_t^2 u_+)(X) + (A_+ u_+)(X) &= 0, & X &\in G^+, \\
B_-(\partial_t^2 u_-)(X) + (A_- u_-)(X) &= 0, & X &\in G^-, \\
u_+(x,0+) = (\partial_t u_+)(x,0+) &= 0, & x &\in S^+, \\
u_-(x,0+) = (\partial_t u_-)(x,0+) &= 0, & x &\in S^-, \\
u_+^+(X) - u_-^-(X) &= f(X), & X &\in \Gamma, \\
(T_+ u_+)^+(X) - (T_- u_-)^-(X) &= g(X), & X &\in \Gamma.
\end{aligned}
$$

(5.1)

Let $v \in C_0^\infty(\mathbb{R}^2 \times [0, \infty))$, and let $v_\pm = \pi^\pm v$. Multiplying the equations of motion for the two plates by v_+^* and v_-^*, integrating over G^+ and G^-, respectively, and adding the resulting equalities, we arrive at the variational equation

$$\int_0^\infty \big[a_+(u_+, v_+) + a_-(u_-, v_-)$$
$$- (B_+^{1/2} \partial_t u_+, B_+^{1/2} \partial_t v_+)_{0;S^+} - (B_-^{1/2} \partial_t u_-, B_-^{1/2} \partial_t v_-)_{0;S^-} \big] \, dt$$
$$= \int_0^\infty (g, v_0)_{0;\partial S} \, dt \quad \forall v \in C_0^\infty(\mathbb{R}^2 \times [0, \infty)), \quad (5.2)$$

where $v_0 = v_+^+ = v_-^-$. Equality (5.2) suggests that the variational version of problem (DT) in the more general case of the nonhomogeneous equation of motion should consist in finding $u = \{u_+, u_-\}$, $u_\pm \in H_{1,0,\kappa}(G^\pm)$, such that

$$\int_0^\infty \big[a_+(u_+, v_+) + a_-(u_-, v_-)$$
$$- (B_+^{1/2} \partial_t u_+, B_+^{1/2} \partial_t v_+)_{0;S^+} - (B_-^{1/2} \partial_t u_-, B_-^{1/2} \partial_t v_-)_{0;S^-} \big] \, dt$$
$$= \int_0^\infty \big[(q, v)_0 + (g, v_0)_{0;\partial S} \big] \, dt \quad \forall v \in C_0^\infty(\mathbb{R}^2 \times [0, \infty)), \qquad (5.3)$$
$$\gamma^+ u_+ - \gamma^- u_- = f,$$

where $q \in H_{-1,k,\kappa}(\mathbb{R}^2 \times (0, \infty))$, $f \in H_{1/2,k,\kappa}(\Gamma)$, and $g \in H_{-1/2,k,\kappa}(\Gamma)$ are prescribed.

5.1 Theorem. *For any* $q \in H_{-1,1,\kappa}(\mathbb{R}^2 \times (0, \infty))$, $f \in H_{1/2,1,\kappa}(\Gamma)$, *and* $g \in H_{-1/2,1,\kappa}(\Gamma)$, $\kappa > 0$, *problem* (5.3) *has a unique solution* $u = \{u_+, u_-\}$, *where* $u_\pm \in H_{1,0,\kappa}(G^\pm)$. *If* $q \in H_{-1,k,\kappa}(\mathbb{R}^2 \times (0, \infty))$, $f \in H_{1/2,k,\kappa}(\Gamma)$, *and* $g \in H_{-1/2,k,\kappa}(\Gamma)$, $k \in \mathbb{R}$, *then* $u_\pm \in H_{1,k-1,\kappa}(G^\pm)$ *and*

$$\|u_+\|_{1,k-1,\kappa;G^+} + \|u_-\|_{1,k-1,\kappa;G^-}$$
$$\leq c(\|q\|_{-1,k,\kappa;\mathbb{R}^2 \times (0,\infty)} + \|f\|_{1/2,k,\kappa;\Gamma} + \|g\|_{-1/2,k,\kappa;\Gamma}). \quad (5.4)$$

Proof. We begin by rewriting (5.3) in terms of Laplace transforms. Then, with the usual notation

$$\hat{u} = \mathcal{L}u, \quad \hat{q} = \mathcal{L}q, \quad \hat{f} = \mathcal{L}f, \quad \hat{g} = \mathcal{L}g,$$

(DT) turns into the problem of finding $\hat{u} \in H_{1,p}(\mathbb{R}^2)$, where

$$\hat{u}(x, p) = \{\hat{u}_+(x, p), \hat{u}_-(x, p)\}, \quad \hat{u}_\pm \in H_{1,p}(S^\pm), \quad p \in \mathbb{C}_\kappa,$$

such that

$$a_+(\hat{u}_+, v_+) + a_-(\hat{u}_-, v_-)$$
$$+ p^2(B_+^{1/2}\hat{u}_+, B_+^{1/2}v_+)_{0;S^+} + p^2(B_-^{1/2}\hat{u}_-, B_-^{1/2}v_-)_{0;S^-}$$
$$= (\hat{q}, v)_0 + (\hat{g}, \gamma v)_{0;\partial S} \quad \forall v \in H_{1,p}(\mathbb{R}^2), \tag{5.5}$$
$$\gamma^+\hat{u}_+ - \gamma^-\hat{u}_- = \hat{f}.$$

Here, $v_\pm = \pi^\pm v$ and γv is the trace of v on ∂S.

First, we consider the case $\hat{f} = 0$. Then (5.5) reduces to the problem

$$a_+(\hat{u}_+, v_+) + a_-(\hat{u}_-, v_-)$$
$$+ p^2(B_+^{1/2}\hat{u}_+, B_+^{1/2}v_+)_{0;S^+} + p^2(B_-^{1/2}\hat{u}_-, B_-^{1/2}v_-)_{0;S^-}$$
$$= (\hat{q}, v)_0 + (\hat{g}, \gamma v)_{0;\partial S} \quad \forall v \in H_{1,p}(\mathbb{R}^2). \tag{5.6}$$

Since $\hat{g} \in H_{-1/2,p}(\partial S)$, the form $(\hat{g}, \gamma v)_{0;\partial S}$ defines a bounded antilinear (conjugate linear) functional on $H_{1,p}(\mathbb{R}^2)$; hence, it can be expressed as

$$(\hat{g}, \gamma v)_{0;\partial S} = (\tilde{q}, v)_0 \quad \forall v \in H_{1,p}(\mathbb{R}^2),$$

where $\tilde{q} \in H_{-1,p}(\mathbb{R}^2)$ and

$$\|\tilde{q}\|_{-1,p} \leq c\|\hat{g}\|_{-1/2,p;\partial S}.$$

Writing $Q = \hat{q} + \tilde{q}$, we arrive at the problem

$$a_+(\hat{u}_+, v_+) + a_-(\hat{u}_-, v_-)$$
$$+ p^2(B_+^{1/2}\hat{u}_+, B_+^{1/2}v_+)_{0;S^+} + p^2(B_-^{1/2}\hat{u}_-, B_-^{1/2}v_-)_{0;S^-}$$
$$= (Q, v)_0 \quad \forall v \in H_{1,p}(\mathbb{R}^2). \tag{5.7}$$

Next, we introduce a bilinear form $a_\kappa(u, v)$ on $[H_{1,p}(\mathbb{R}^2)]^2$ by

$$a_\kappa(u, v) = \tfrac{1}{2}\kappa^2\big[(B_+^{1/2}u_+, B_+^{1/2}v_+)_{0;S^+} + (B_-^{1/2}u_-, B_-^{1/2}v_-)_{0;S^-}\big]$$
$$+ a_+(u_+, v_+) + a_-(u_-, v_-).$$

It is obvious that $a_\kappa(u, v)$ is continuous and coercive on $[H_{1,p}(\mathbb{R}^2)]^2$; therefore, it generates a self-adjoint nonnegative operator $\mathcal{A}_\kappa$ through the equality

$$(\mathcal{A}_\kappa u, v)_0 = a_\kappa(u, v) \quad \forall u, v \in H_{1,p}(\mathbb{R}^2).$$

$\mathcal{A}_\kappa$ is a homeomorphism from $H_{1,p}(\mathbb{R}^2)$ to $H_{-1,p}(\mathbb{R}^2)$. We can now write (5.7) in the form

$$\mathcal{A}_\kappa \hat{u} + \big(p^2 - \tfrac{1}{2}\kappa^2\big)\tilde{B}\hat{u} = Q, \tag{5.8}$$

where $\tilde{B}\hat{u} = (B_+\hat{u}_+, B_-\hat{u}_-) \in H_{-1,p}(\mathbb{R}^2)$. In turn, (5.8) can be written as

$$\hat{u} + \left(p^2 - \tfrac{1}{2}\kappa^2\right)\mathcal{A}_\kappa^{-1}\tilde{B}\hat{u} = \mathcal{A}_\kappa^{-1}Q,$$

or

$$u_b + \left(p^2 - \tfrac{1}{2}\kappa^2\right)\mathcal{B}_\kappa u_b = \tilde{B}^{1/2}\mathcal{A}_\kappa^{-1}Q, \tag{5.9}$$

where $u_b = \tilde{B}^{1/2}\hat{u}$ and $\mathcal{B}_\kappa = \tilde{B}^{1/2}\mathcal{A}_\kappa^{-1}\tilde{B}^{1/2}$. Following the scheme used in the proof of Theorem 2.2, we show that (5.9), regarded as an equation in $H_{1,p}(\mathbb{R}^2)$, is equivalent to itself in $L^2(\mathbb{R}^2)$, and that $\mathcal{B}_\kappa$ is self-adjoint and nonnegative on $L^2(\mathbb{R}^2)$. Consequently, (5.9) is uniquely solvable, so (5.7)—hence, also (5.6)—have a unique solution $\hat{u} \in H_{1,p}(\mathbb{R}^2)$ for any $Q \in H_{-1,p}(\mathbb{R}^2)$.

Taking $v = \hat{u}$ in (5.7) and separating the real and imaginary parts, we obtain

$$\|\hat{u}\|_{1,p}^2 \le c|p||(Q,\hat{u})_0|;$$

therefore,

$$\|\hat{u}\|_{1,p} \le c|p|\|Q\|_{-1,p} \le c|p|(\|\hat{q}\|_{-1,p} + \|\hat{g}\|_{-1/2,p;\partial S}). \tag{5.10}$$

In the general case we construct

$$w_+ = l^+\hat{f} \in H_{1,p}(S^+), \quad w = \{w_+, 0\},$$

then seek a solution of (5.5) of the form $\hat{u} = w + \tilde{u}$. Clearly, $\tilde{u} \in H_{1,p}(\mathbb{R}^2)$ satisfies

$$
\begin{aligned}
a_+(\tilde{u}_+,&v_+) + a_-(\tilde{u}_-,v_-) \\
&+ p^2(B_+^{1/2}\tilde{u}_+, B_+^{1/2}v_+)_{0;S^+} + p^2(B_-^{1/2}\tilde{u}_-, B_-^{1/2}v_-)_{0;S^-} \\
&= (\hat{q},v)_0 + (\hat{g},\gamma v)_{0;\partial S} - a_+(w_+,v_+) \\
&\quad - p^2(B_+^{1/2}w_+, B_+^{1/2}v_+)_{0;S^+} \quad \forall v \in H_{1,p}(\mathbb{R}^2). \tag{5.11}
\end{aligned}
$$

Since the extension operator l^+ is continuous (uniformly with respect to the parameter $p \in \mathbb{C}$ from $H_{1/2,p}(\partial S)$ to $H_{1,p}(S^+)$, we have

$$\|l^+\hat{f}\|_{1,p;S^+} \le c\|\hat{f}\|_{1/2,p;\partial S},$$

which leads to

$$|a_+(w_+,v_+) + p^2(B_+^{1/2}w_+, B_+^{1/2}v_+)_{0;S^+}| \le c\|w_+\|_{1,p;S^+}\|v_+\|_{1,p;S^+}$$

$$\le c\|\hat{f}\|_{1/2,p;\partial S}\|v\|_{1,p}.$$

Therefore, (5.11) has a solution $\tilde{u} \in H_{1,p}(\mathbb{R}^2)$ and, by (5.10),

$$\|\tilde{u}\|_{1,p} \leq c|p|(\|\hat{q}\|_{-1,p} + \|\hat{f}\|_{1/2,p;\partial S} + \|\hat{g}\|_{-1/2,p;\partial S}).$$

This implies that (5.5) also has a solution $\hat{u}$, which satisfies

$$\|\hat{u}_+\|_{1,p;S^+} + \|\hat{u}_-\|_{1,p;S^-}$$
$$\leq c|p|(\|\hat{q}\|_{-1,p} + \|\hat{f}\|_{1/2,p;\partial S} + \|\hat{g}\|_{-1/2,p;\partial S}). \tag{5.12}$$

We now use (5.12) to show successively that $\hat{U} = \{\hat{U}_+, \hat{U}_-\}$, where $\hat{U}_\pm(p) = \hat{u}_\pm(\cdot, p)$, is bounded, continuous, and holomorphic from $\mathbb{C}_\kappa$ to $H_1(S^+) \times H_1(S^-)$. This and (5.12) prove that problem (DT) has a solution, which satisfies estimate (5.4).

To show that this solution is unique, suppose that $u \in H_{1,0,\kappa}(\mathbb{R}^2 \times (0, \infty))$ satisfies

$$\int_0^\infty [a_+(u_+, v_+) + a_-(u_-, v_-) - (B_+^{1/2}\partial_t u_+, B_+^{1/2}\partial_t v_+)_{0;S^+}$$
$$- (B_-^{1/2}\partial_t u_-, B_-^{1/2}\partial_t v_-)_{0;S^-}]\, dt = 0 \quad \forall v \in C_0^\infty(\mathbb{R}^2 \times [0, \infty)).$$

Repeating the relevant part of the proof of Theorem 2.3 with $L^2(S^+)$, $\mathring{H}_1(S^+)$, and $a_+(u, v)$ replaced by $L^2(\mathbb{R}^2)$, $H_1(\mathbb{R}^2)$, and $a_+(u_+, v_+) + a_-(u_-, v_-)$, we arrive at the desired conclusion. $\quad\square$

Having established the existence of a unique weak solution to our problem, we turn our attention to the representation of this solution in terms of various combinations of time-dependent (retarded) single-layer and double-layer potentials and prove the unique solvability of the corresponding systems of nonstationary boundary integral equations.

Let $V_\pm\alpha_\pm$ and $W_\pm\beta_\pm$ be the retarded single-layer and double-layer potentials constructed for the plates corresponding to $S^\pm$, respectively. The boundary operators generated by these potentials are denoted by $V_{+,0}$, $V_{-,0}$, W_+^+, W_-^-, and N_+, N_-. The Poincaré–Steklov operators for $S^\pm$ are denoted by $\mathcal{T}^\pm$. Similar notation is used for their Laplace transform counterparts, with the only difference that the latter carry a subscript p.

First we represent the (weak) solution $u = \{u_+, u_-\}$ of (5.1) in the form

$$u_+(X) = (V_+\alpha_+)(X), \quad X \in G^+,$$
$$u_-(X) = (V_-\alpha_-)(X), \quad X \in G^-, \tag{5.13}$$

where $\alpha_\pm$ are unknown densities defined on Γ. This yields the system of nonstationary boundary equations

$$V_{+,0}\alpha_+ - V_{-,0}\alpha_- = f,$$
$$\mathcal{T}_+^+ V_{+,0}\alpha_+ - \mathcal{T}_-^- V_{-,0}\alpha_- = g. \tag{5.14}$$

The second representation is

$$u_+(X) = (W_+\beta_+)(X), \quad X \in G^+,$$
$$u_-(X) = (W_-\beta_-)(X), \quad X \in G^-,$$
(5.15)

where $\beta_\pm$ are unknown densities defined on Γ. Representation (5.15) yields the system of boundary integral equations

$$W_+^+\beta_+ - W_-^-\beta_- = f,$$
$$N_+\beta_+ - N_-\beta_- = g.$$
(5.16)

In the third case we seek the solution as

$$u_+(X) = (V_+\alpha_+)(X), \quad X \in G^+,$$
$$u_-(X) = (W_-\beta_-)(X), \quad X \in G^-,$$
(5.17)

which leads to the system

$$V_{+,0}\alpha_+ - W_-^-\beta_- = f,$$
$$T_+^+V_{+,0}\alpha_+ - N_-\beta = g.$$
(5.18)

Finally, seeking the solution in the form

$$u_+(X) = (W_+\beta_+)(X), \quad X \in G^+,$$
$$u_-(X) = (V_-\alpha_-)(X), \quad X \in G^-,$$
(5.19)

we arrive at the system

$$W_+^+\beta_+ - V_{-,0}\alpha_- = f,$$
$$N_+\beta_+ - T_-^-V_{-,0}\alpha_- = g.$$
(5.20)

5.2 Theorem. *For any $f \in H_{1/2,k,\kappa}(\Gamma)$ and $g \in H_{-1/2,k,\kappa}(\Gamma)$, $\kappa > 0$, systems (5.14), (5.16), (5.18), and (5.20) have unique solutions*

$$\alpha_\pm \in H_{-1/2,k-2,\kappa}(\Gamma), \quad \beta_\pm \in H_{1/2,k-2,\kappa}(\Gamma).$$

In each case, the corresponding function u defined by (5.13), (5.15), (5.17), or (5.19) belongs to $H_{1,k-1,\kappa}(G^+) \times H_{1,k-1,\kappa}(G^-)$. If $k \geq 1$, then u is the (weak) solution of problem (DT).

Proof. We rewrite all the systems of boundary equations in terms of their Laplace transforms. For $p \in \mathbb{C}_\kappa$, systems (5.14), (5.16), (5.18), and (5.20) take, respectively, the form

$$V_{p,+,0}\hat{\alpha}_+ - V_{p,-,0}\hat{\alpha}_- = \hat{f},$$
$$\mathcal{T}_{p,+}^+ V_{p,+,0}\hat{\alpha}_+ - \mathcal{T}_{p,-}^- V_{p,-,0}\hat{\alpha}_- = \hat{g}, \tag{5.21}$$

$$W_{p,+}^+ \hat{\beta}_+ - W_{p,-}^- \hat{\beta}_- = \hat{f},$$
$$N_{p,+}\hat{\beta}_+ - N_{p,-}\hat{\beta}_- = \hat{g}, \tag{5.22}$$

$$V_{p,+,0}\hat{\alpha}_+ - W_{p,-}^- \hat{\beta}_- = \hat{f},$$
$$\mathcal{T}_{p,+}^+ V_{p,+,0}\hat{\alpha}_+ - N_{p,-}\hat{\beta}_- = \hat{g}, \tag{5.23}$$

$$W_{p,+}^+ \hat{\beta}_+ - V_{p,-,0}\hat{\alpha}_- = \hat{f},$$
$$N_{p,+}\hat{\beta}_+ - \mathcal{T}_{p,-}^- V_{p,-,0}\hat{\alpha}_- = \hat{g}. \tag{5.24}$$

Next, we consider the solution $\hat{u}(x,p) = \{\hat{u}_+(x,p), \hat{u}_-(x,p)\}$ of problem (5.5) for any $p \in \mathbb{C}_\kappa$. In §5.1 it was shown that

$$\|\hat{u}_+\|_{1,p;S^+} + \|\hat{u}_-\|_{1,p;S^-} \le c|p|(\|\hat{f}\|_{1/2,p;\partial S} + \|\hat{g}\|_{-1/2,p;\partial S}). \tag{5.25}$$

We write $\hat{f}_\pm = \gamma^\pm \hat{u}_\pm$. By the trace theorem and (5.25),

$$\|\hat{f}_+\|_{1/2,p;\partial S} + \|\hat{f}_-\|_{1/2,p;\partial S}$$
$$\le c|p|(\|\hat{f}\|_{1/2,p;\partial S} + \|\hat{g}\|_{-1/2,p;\partial S}). \tag{5.26}$$

We also write $\hat{g}_+ = \mathcal{T}_{p,+}^+ \hat{f}_+$ and $\hat{g}_- = \mathcal{T}_{p,-}^- \hat{f}_-$. By (5.25) and (3.5),

$$\|\hat{g}_+\|_{-1/2,p;\partial S} + \|\hat{g}_-\|_{-1/2,p;\partial S}$$
$$\le c(\|\hat{u}_+\|_{1,p;S^+} + \|\hat{u}_-\|_{1,p;S^-})$$
$$\le c|p|(\|\hat{f}\|_{1/2,p;\partial S} + \|\hat{g}\|_{-1/2,p;\partial S}). \tag{5.27}$$

We now consider system (5.21). Let $\hat{\alpha}_\pm = V_{p,\pm}^{-1}\hat{f}_\pm$. Then (5.26) shows that

$$\|\hat{\alpha}_+\|_{-1/2,p;\partial S} + \|\hat{\alpha}_-\|_{-1/2,p;\partial S}$$
$$\le c|p|(\|\hat{f}_+\|_{1/2,p;\partial S} + \|\hat{f}_-\|_{1/2,p;\partial S})$$
$$\le c|p|^2(\|\hat{f}\|_{1/2,p;\partial S} + \|\hat{g}\|_{-1/2,p;\partial S}). \tag{5.28}$$

Clearly, $\hat{u}_+ = V_{p,+}\hat{\alpha}_+$ and $\hat{u}_- = V_{p,-}\hat{\alpha}_-$ (hence, also $\hat{\alpha}_+$ and $\hat{\alpha}_-$) satisfy (5.21). The assertion concerning (5.14) follows from (5.28) and Theorem 5.1.

In the case of (5.22), we take $\hat\beta_+ = N_{p,+}^{-1}\hat{g}_+$ and $\hat\beta_- = N_{p,-}^{-1}\hat{g}_-$. From the properties of the operators $(N_{p,\pm})^{-1}$ and (5.27) it follows that

$$\|\hat\beta_+\|_{1/2,p;\partial S} + \|\hat\beta_-\|_{1/2,p;\partial S}$$

$$\leq c|p|\big(\|\hat{g}_+\|_{-1/2,p;\partial S} + \|\hat{g}_-\|_{-1/2,p;\partial S}\big)$$

$$\leq c|p|^2\big(\|\hat{f}\|_{1/2,p;\partial S} + \|\hat{g}\|_{-1/2,p;\partial S}\big). \qquad (5.29)$$

The desired statement now follows from (5.29) and Theorem 5.1.

For systems (5.23) and (5.24) we set

$$\hat\alpha_+ = V_{p,+,0}^{-1}\hat{f}_+, \quad \hat\beta_- = N_{p,-}^{-1}\hat{g}_-,$$

$$\hat\beta_+ = N_{p,+}^{-1}\hat{g}_+, \quad \hat\alpha_- = V_{p,-,0}^{-1}\hat{f}_-,$$

respectively. The proof is completed by repeating the above arguments, with the obvious changes. $\qquad\square$

5.2 Multiply Connected Finite Plate

Consider a plate with Lamé constants λ_+, μ_+ and density ρ_+, and characterized by the thickness parameter h_+ and a middle plane domain S^+ whose boundary ∂S consists of two simple closed C^2-curves ∂S_1 and ∂S_2 such that ∂S_1 lies strictly inside the finite domain enclosed by ∂S_2. The unit normals on ∂S_1 and ∂S_2 are directed outwards with respect to S^+. Let S_i^- be the domain interior to ∂S_1 and S_e^- the domain exterior to ∂S_2.

For $k \in \mathbb{R}$, $\kappa > 0$, and $p \in \mathbb{C}_0$, we define the spaces

$$H_{\pm 1/2,p}(\partial S) = H_{\pm 1/2,p}(\partial S_1) \times H_{\pm 1/2,p}(\partial S_2),$$

$$H_{\pm 1/2,k,\kappa}(\partial S) = H_{\pm 1/2,k,\kappa}(\partial S_1) \times H_{\pm 1/2,k,\kappa}(\partial S_2),$$

$$H_{\pm 1/2,k,\kappa}(\Gamma) = H_{\pm 1/2,k,\kappa}(\Gamma_1) \times H_{\pm 1/2,k,\kappa}(\Gamma_2),$$

where

$$\Gamma_\nu = \partial S_\nu \times (0,\infty), \quad \nu = 1,2, \quad \Gamma = \Gamma_1 \cup \Gamma_2.$$

We also write

$$L^2(\partial S) = L^2(\partial S_1) \times L^2(\partial S_2).$$

If $f = \{f_1, f_2\}$ and $\varphi = \{\varphi_1, \varphi_2\}$ are elements of $\big[L^2(\partial S)\big]^2$, then their $L^2(\partial S)$-inner product is defined as

$$(f,\varphi)_{0;\partial S} = (f_1, \varphi_1)_{0;\partial S_1} + (f_2, \varphi_2)_{0;\partial S_2}.$$

We denote by $\gamma^+ = \{\gamma_1^+, \gamma_2^+\}$ the trace operator that maps $H_{1,p}(S^+)$ continuously onto $H_{1/2,p}(\partial S)$. Thus, for $u \in H_{1,p}(S^+)$, we write $\gamma^+ u = \{\gamma_1^+ u, \gamma_2^+ u\}$, where $\gamma_\nu^+ u$ are the traces of u on ∂S_ν, $\nu = 1, 2$. According to the notational convention adopted in §1.1, if u is a function in a space of originals, then its trace on Γ is also denoted by $\gamma^+ u = (\gamma_1^+ u, \gamma_2^+ u)$. For any $k \in \mathbb{R}$, the trace operator

$$\gamma^+ : H_{1,k,\kappa}(G^+) \to H_{1/2,k,\kappa}(\Gamma)$$

is continuous. Finally, we introduce extension operators l_ν^+, $\nu = 1, 2$, such that for any $f_\nu \in H_{1/2,p}(\partial S_\nu)$,

$$l_\nu^+ f_\nu \in H_{1,p}(S^+), \quad \gamma_{3-\nu}^+ l_\nu^+ f_\nu = 0, \quad \nu = 1, 2 \text{ (not summed)}.$$

If $f = \{f_1, f_2\} \in H_{1/2,p}(\partial S)$, then we set $l^+ f = l_1^+ f_1 + l_2^+ f_2$. It is obvious that $\gamma^+ l^+ f = f$ and that l^+ is continuous (uniformly with respect to p) from $H_{1/2,p}(\partial S)$ to $H_{1,p}(S^+)$.

We now formulate three initial-boundary value problems for the plate with middle section S^+. The classical problem (DMCD) consists in finding $u \in C^2(G^+) \cap C^1(\bar{G}^+)$ such that

$$B_+(\partial_t^2 u)(X) + (A_+ u)(X) = 0, \quad X \in G^+,$$
$$u(x, 0+) = (\partial_t u)(x, 0+) = 0, \quad x \in S^+,$$
$$u^+(X) = f_1(X), \quad X \in \Gamma_1,$$
$$u^+(X) = f_2(X), \quad X \in \Gamma_2,$$

where f_1 and f_2 are prescribed. Consequently, in the variational version of (DMCD) we seek $u \in H_{1,0,\kappa}(G^+)$ satisfying

$$\int_0^\infty \left[a_+(u, v) - (B_+^{1/2} \partial_t u, B_+^{1/2} \partial_t v)_{0;S^+} \right] dt = 0$$

$$\forall v \in C_0^\infty(\bar{G}^+), \ v^+ = 0,$$
$$\gamma^+ u = f,$$

where a_+ and B_+ have the obvious meaning.

The classical problem (DMCN) consists in finding $u \in C^2(G^+) \cap C^1(\bar{G}^+)$ such that

$$B_+(\partial_t^2 u)(X) + (A_+ u)(X) = 0, \quad X \in G^+,$$
$$u(x, 0+) = (\partial_t u)(x, 0+) = 0, \quad x \in S^+,$$
$$(T_+ u)^+(X) = g_1(X), \quad X \in \Gamma_1,$$
$$(T_+ u)^+(X) = g_2(X), \quad X \in \Gamma_2,$$

where g_1 and g_2 are prescribed. In its corresponding variational version, we seek $u \in H_{1,0,\kappa}(G^+)$ satisfying

$$\int\limits_0^\infty \left[a_+(u,v) - (B_+^{1/2}\partial_t u, B_+^{1/2}\partial_t v)_{0;S^+}\right] dt$$
$$= \int\limits_0^\infty (g, v^+)_{0;\partial S}\, dt \quad \forall v \in C_0^\infty(\bar{G}^+),$$

where $v^+ = \{v_1^+, v_2^+\}$ and $g = \{g_1, g_2\}$.

The classical problem (DMCM) consists in finding $u \in C^2(G^+) \cap C^1(\bar{G}^+)$ such that

$$B_+(\partial_t^2 u)(X) + (A_+ u)(X) = 0, \quad X \in G^+,$$
$$u(x, 0+) = (\partial_t u)(x, 0+) = 0, \quad x \in S^+,$$
$$(T_+ u)^+(X) = g_1(X), \quad X \in \Gamma_1,$$
$$u^+(X) = f_2(X), \quad X \in \Gamma_2,$$

where g_1 and f_2 are prescribed. In the variational version of this problem, we seek $u \in H_{1,0,\kappa}(G^+)$ satisfying

$$\int\limits_0^\infty \left[a_+(u,v) - (B_+^{1/2}\partial_t u, B_+^{1/2}\partial_t v_1^+)_{0;S^+}\right] dt$$
$$= \int\limits_0^\infty (g_1, v_1^+)_{0;\partial S_1}\, dt \quad \forall v \in C_0^\infty(\bar{G}^+), \ v_2^+ = 0,$$
$$\gamma_2^+ u = f_2.$$

5.3 Theorem. *For any $f_\nu \in H_{1/2,1,\kappa}(\Gamma_\nu)$ and $g_\nu \in H_{-1/2,1,\kappa}(\Gamma_\nu)$, $\nu = 1, 2$, $\kappa > 0$, problems (DMCD), (DMCN), and (DMCM) have unique solutions $u \in H_{1,0,\kappa}(G^+)$. If $f_\nu \in H_{1/2,k,\kappa}(\Gamma_\nu)$ and $g_\nu \in H_{-1/2,k,\kappa}(\Gamma_\nu)$, $k \in \mathbb{R}$, then each of these solutions u satisfies the corresponding estimate*

$$\|u\|_{1,k-1,\kappa;G^+} \le c\big(\|f_1\|_{1/2,k,\kappa;\Gamma_1} + \|f_2\|_{1/2,k,\kappa;\Gamma_2}\big),$$
$$\|u\|_{1,k-1,\kappa;G^+} \le c\big(\|g_1\|_{-1/2,k,\kappa;\Gamma_1} + \|g_2\|_{-1/2,k,\kappa;\Gamma_2}\big),$$
$$\|u\|_{1,k-1,\kappa;G^+} \le c\big(\|g_1\|_{-1/2,k,\kappa;\Gamma_1} + \|f_2\|_{1/2,k,\kappa;\Gamma_2}\big).$$

Proof. Going over to Laplace transforms in problems (DMCD), (DMCN), and (DMCM), we arrive at new problems for every $p \in \mathbb{C}_\kappa$. Thus, in (MCD$_p$), we seek $\hat{u} \in H_{1,p}(S^+)$ such that

$$p^2(B_+^{1/2}\hat{u}, B_+^{1/2}v)_{0;S^+} + a_+(\hat{u}, v) = 0 \quad \forall v \in \mathring{H}_{1,p}(S^+),$$
$$\gamma^+ \hat{u} = \hat{f}. \tag{5.30}$$

In (MCN$_p$), we seek $\hat{u} \in H_{1,p}(S^+)$ such that

$$p^2(B_+^{1/2}\hat{u}, B_+^{1/2}v)_{0;S^+} + a_+(\hat{u}, v)$$
$$= (\hat{g}, \gamma^+v)_{0;\partial S} \quad \forall v \in H_{1,p}(S^+), \tag{5.31}$$

where $\gamma^+v = \{\gamma_1^+v, \gamma_2^+v\}$ is the trace of v on ∂S.

Finally, in (MCM$_p$), we seek $\hat{u} \in H_{1,p}(S^+)$ such that

$$p^2(B_+^{1/2}\hat{u}, B_+^{1/2}v)_{0;S^+} + a_+(\hat{u}, v)$$
$$= (\hat{g}_1, \gamma_1^+v)_{0;\partial S_1} \quad \forall v \in \mathring{H}_{1,p}(S^+, \partial S_1), \tag{5.32}$$
$$\gamma_2^+\hat{u} = \hat{f}_2,$$

where $\mathring{H}_{1,p}(S^+, \partial S_1)$ is the subspace of all u in $H_{1,p}(S^+)$ such that $\gamma_2^+u = 0$.

We represent the solution of (MCD$_p$) in the form $\hat{u} = w + l^+\hat{f}$. It is obvious that $w \in \mathring{H}_{1,p}(S^+)$ satisfies

$$p^2(B_+^{1/2}w, B_+^{1/2}v)_{0;S^+} + a_+(w, v)$$
$$= (Q_1, v)_{0;S^+} \quad \forall v \in \mathring{H}_{1,p}(S^+), \tag{5.33}$$

where $Q_1 \in H_{-1,p}(S^+)$ is defined for $v \in \mathring{H}_{1,p}(S^+)$ by

$$(Q_1, v)_{0;S^+} = -\left[p^2(B_+^{1/2}l^+\hat{f}, B_+^{1/2}v)_{0;S^+} + a_+(l^+\hat{f}, v)\right].$$

We see that

$$|(Q_1, v)_{0;S^+}| \le c\|l^+\hat{f}\|_{1,p;S^+}\|v\|_{1,p} \le c\|\hat{f}\|_{1/2,p;\partial S}\|v\|_{1,p};$$

hence,

$$\|Q_1\|_{-1,p;S^+} \le c\|\hat{f}\|_{1/2,p;\partial S}.$$

The unique solvability of (5.33) and the estimates

$$\|w\|_{1,p} \le c|p|\|Q_1\|_{-1,p;S^+} \le c|p|\|\hat{f}\|_{1/2,p;\partial S}$$

are established by the method used in the proof of Theorem 2.1. Therefore, (5.30) is uniquely solvable in $H_{1,p}(S^-)$ and

$$\|\hat{u}\|_{1,p;S^+} \le c|p|\|\hat{f}\|_{1/2,p;\partial S}. \tag{5.34}$$

In the case of equation (5.31) we remark that for $\hat{g} \in H_{-1/2,p}(\partial S)$, the form $(\hat{g}, \gamma^+v)_{0;\partial S}$ defines a bounded antilinear (conjugate linear) functional on the space $H_{1,p}(S^+)$, so it can be written as

$$(\hat{g}, \gamma^+v)_{0;\partial S} = (Q_2, v)_{0;S^+} \quad \forall v \in H_{1,p}(S^+).$$

Since

$$|(Q_2, v)_{0;S^+}| \le c\|\hat{g}\|_{-1/2,p;\partial S} \|\gamma^+ v\|_{1/2,p;\partial S}$$

$$\le c\|\hat{g}\|_{-1/2,p;\partial S} \|v\|_{1,p;S^+},$$

it follows that $Q_2 \in \mathring{H}_{-1,p}(S^+)$ satisfies

$$\|Q_2\|_{-1,p} \le c\|\hat{g}\|_{-1/2,p;\partial S}.$$

We then write (5.31) in the form

$$a_{+,\kappa}(\hat{u}, v) + (p^2 - \tfrac{1}{2}\kappa^2)(B_+^{1/2}\hat{u}, B_+^{1/2}v)_{0;S^+}$$

$$= (Q_2, v)_{0;S^+} \quad \forall v \in H_{1,p}(S^+), \qquad (5.35)$$

where the bilinear form $a_{+,\kappa}$ is defined on $\left[H_{1,p}(S^+)\right]^2$ by

$$a_{+,\kappa}(u, v) = \tfrac{1}{2}\kappa^2(B_+^{1/2}u, B_+^{1/2}v)_{0;S^+} + a_+(u, v).$$

It is obvious that $a_{+,\kappa}$ is continuous and coercive on $\left[H_{1,p}(S^+)\right]^2$; hence, it generates an operator $\mathcal{A}_\kappa$ that is a homeomorphism from $H_{1,p}(S^+)$ to $\mathring{H}_{-1,p}(S^+)$, by means of the equality

$$(\mathcal{A}_\kappa u, v)_{0;S^+} = a_{+,\kappa}(u, v) \quad \forall u, v \in H_{1,p}(S^+).$$

We now rewrite (5.35) in the form

$$\mathcal{A}_\kappa \hat{u} + \left(p^2 - \tfrac{1}{2}\kappa^2\right)B_+\hat{u} = Q_2. \qquad (5.36)$$

Applying the Fredholm Alternative, we prove the unique solvability of (5.36)—hence, of (5.31)—in $H_{1,p}(S^+)$ for any $\hat{g} \in H_{-1/2,p}(\partial S)$, $p \in \mathbb{C}_\kappa$. Separating the real and imaginary parts in (5.35), we obtain

$$\|\hat{u}\|_{1,p;S^+} \le c|p|\|Q_2\|_{-1,p} \le c|p|\|\hat{g}\|_{-1/2,p;\partial S}. \qquad (5.37)$$

The solution of (5.32) is sought in the form $\hat{u} = w + l_2^+ \hat{f}_2$, where $w \in \mathring{H}_{1,p}(S^+, \partial S_1)$. Let $\mathring{H}_{-1,p}(S^+, \partial S_2)$ be the dual of $\mathring{H}_{1,p}(S^+, \partial S_1)$ with respect to the duality generated by the inner product in $L^2(S^+)$. It is obvious that $\mathring{H}_{-1,p}(S^+, \partial S_2)$ is the subspace of $H_{-1,p}(S^+ \cup \bar{S}_i^-)$ that consists of all q with $\mathrm{supp}\, q \subset \bar{S}^+$. We denote the norm of $q \in \mathring{H}_{-1,p}(S^+, \partial S_2)$ by $\|q\|_{-1,p;S^+ \cup \bar{S}_i^-}$.

We see that $(\hat{g}_1, \gamma_1^+ v)_{0;\partial S_1}$ defines a bounded antilinear (conjugate linear) functional on $\mathring{H}_{1,p}(S^+, \partial S_1)$; hence, it can be expressed as

$$(\hat{g}_1, \gamma_1^+ v)_{0;\partial S_1} = (\tilde{Q}_3, v)_{0;S^+} \quad \forall v \in \mathring{H}_{1,p}(S^+, \partial S_1),$$

where $\tilde{Q}_3 \in \mathring{H}_{-1,p}(S^+, \partial S_2)$ and

$$\|\tilde{Q}_3\|_{-1,p;S^+\cup\bar{S}_i^-} \leq c\|\hat{g}_1\|_{-1/2,p;\partial S_1}.$$

We write (5.32) as

$$p^2(B_+^{1/2}w, B_+^{1/2}v)_{0;S^+} + a_+(w,v)$$
$$= (\tilde{Q}_3, v)_{0;S^+} - p^2(B_+^{1/2}l_2^+\hat{f}_2, B_+^{1/2}v)_{0;S^+} - a_+(l_2^+\hat{f}_2, v)$$
$$\forall v \in \mathring{H}_{1,p}(S^+, \partial S_1). \qquad (5.38)$$

Since

$$p^2(B_+^{1/2}l_2^+\hat{f}_2, B_+^{1/2}v)_{0;S^+} + a_+(l_2^+\hat{f}_2, v) = (q, v)_{0;S^+},$$

where $q \in H_{-1,p}(S^+, \partial S_2)$ and

$$\|q\|_{-1,p;S^+\cup\bar{S}_i^-} \leq c\|\hat{f}_2\|_{1/2,p;\partial S_2},$$

we make the notation $\tilde{Q}_3 - q = Q_3 \in \mathring{H}_{-1,p}(S^+, \partial S_2)$ and see that

$$\|Q_3\|_{-1,p;S^+\cup\bar{S}_i^-} \leq c(\|\hat{f}_2\|_{1/2,p;\partial S_2} + \|\hat{g}_1\|_{-1/2,p;\partial S_1}).$$

Then (5.38) becomes

$$p^2(B_+^{1/2}w, B_+^{1/2}v) + a_+(w,v)$$
$$= (Q_3, v)_{0;S^+} \quad \forall v \in \mathring{H}_{1,p}(S^+, \partial S_1). \qquad (5.39)$$

Taking into account the continuity and coerciveness of the form $a_+(u,v)$ on $[\mathring{H}_{1,p}(S^+, \partial S_1)]^2$, we adapt the proof of Theorem 2.1 to our situation and conclude that (5.39) has a unique solution $w \in \mathring{H}_{1,p}(S^+, \partial S_1)$ and that

$$\|w\|_{1,p;S^+\cup\bar{S}_i^-} \leq c|p|(\|\hat{f}_2\|_{1/2,p;\partial S_2} + \|\hat{g}_1\|_{-1/2,p;\partial S_1}).$$

Consequently, (5.32) has a unique solution $\hat{u} \in H_{1,p}(S^+)$ and

$$\|\hat{u}\|_{1,p;S^+} \leq c|p|(\|\hat{f}_2\|_{1/2,p;\partial S_2} + \|\hat{g}_1\|_{-1/2,p;\partial S_1}). \qquad (5.40)$$

To arrive at the desired assertion, we use (5.34), (5.37), and (5.40), and repeat the arguments in the proof of Theorem 2.3 with the obvious changes. □

We now construct an algebra of boundary operators in the case of multiply connected plates. We perform this construction in Sobolev spaces with a parameter; the corresponding assertions in spaces of originals can then be easily formulated. In what follows, we use the notation

$$a_{-,i}(u,v) = \int_{S_i^-} E(u,v)\, dx, \quad a_{-,e}(u,v) = \int_{S_e^-} E(u,v)\, dx,$$

where the bilinear form E is for the plate with the Lamé coefficients λ_+ and μ_+.

Let $f = \{f_1, f_2\}$ and $\varphi = \{\varphi_1, \varphi_2\}$ be any elements of $H_{1/2,p}(\partial S)$, let $u_+ \in H_{1,p}(S^+)$ be the solution of (5.30) with boundary data f, and suppose that $u_-^{(i)} \in H_{1,p}(S_i^-)$ and $u_-^{(e)} \in H_{1,p}(S_e^-)$ satisfy

$$p^2(B_+^{1/2} u_-^{(i)}, B_+^{1/2} v_-^{(i)})_{0;S_i^-} + a_{-,i}(u_-^{(i)}, v_-^{(i)}) = 0 \quad \forall v_-^{(i)} \in \mathring{H}_{1,p}(S_i^-),$$

$$\gamma_1^- u_-^{(i)} = f_1$$

and

$$p^2(B_+^{1/2} u_-^{(e)}, B_+^{1/2} v_-^{(e)}) + a_{-,e}(u_-^{(e)}, v_-^{(e)}) = 0 \quad \forall v_-^{(e)} \in \mathring{H}_{1,p}(S_e^-),$$

$$\gamma_2^- u_-^{(e)} = f_2,$$

where γ_1^- and γ_2^- are the trace operators that map $H_{1,p}(S_i^-)$ and $H_{1,p}(S_e^-)$ continuously onto $H_{1/2,p}(\partial S_1)$ and $H_{1/2,p}(\partial S_2)$, respectively. Also, let $v_-^{(i)}$ and $v_-^{(e)}$ be elements of $H_{1,p}(S_i^-)$ and $H_{1,p}(S_e^-)$ such that $\gamma_1^- v_-^{(i)} = \varphi_1$ and $\gamma_2^- v_-^{(e)} = \varphi_2$. We define the Poincaré–Steklov operators $\mathcal{T}_p^{\pm}$ by

$$(\mathcal{T}_p^+ f, \varphi)_{0;\partial S} = (\mathcal{T}_{p,1}^+ f, \varphi_1)_{0;\partial S_1} + (\mathcal{T}_{p,2}^+ f, \varphi_2)_{0;\partial S_2}$$

$$= p^2(B_+^{1/2} u_+, B_+^{1/2} v_+)_{0;S^+} + a_+(u_+, v_+), \qquad (5.41)$$

where $v_+ \in H_{1,p}(S^+)$ and $\gamma^+ v_+ = \varphi$, and

$$(\mathcal{T}_p^- f, \varphi)_{0;\partial S} = (\mathcal{T}_{p,1}^- f_1, \varphi_1)_{0;\partial S_1} + (\mathcal{T}_{p,2}^- f_2, \varphi_2)_{0;\partial S_2}$$

$$= -\big[p^2(B_+^{1/2} u_-^{(i)}, B_+^{1/2} v_-^{(i)})_{0;S_i^-} + a_{-,i}(u_-^{(i)}, v_-^{(i)})$$

$$+ p^2(B_+^{1/2} u_-^{(e)}, B_+^{1/2} v_-^{(e)})_{0;S_e^-} + a_{-,e}(u_-^{(e)}, v_-^{(e)})\big].$$

5.4 Lemma. *For any $p \in \mathbb{C}_0$, the operators $\mathcal{T}_p^{\pm}$ are homeomorphisms from $H_{1/2,p}(\partial S)$ to $H_{-1/2,p}(\partial S)$, and for any $f \in H_{1/2,p}(\partial S)$, $p \in \bar{\mathbb{C}}_\kappa$, $k > 0$,*

$$\|\mathcal{T}_p^{\pm} f\|_{-1/2,p;\partial S} \le c|p| \|f\|_{1/2,p;\partial S}, \qquad (5.42)$$

$$\|f\|_{1/2,p;\partial S} \le c|p| \|\mathcal{T}_p^{\pm} f\|_{-1/2,p;\partial S}. \qquad (5.43)$$

Proof. The assertion concerning (5.42) and (5.43) for $\mathcal{T}^-$ has already been proved in Lemma 3.1. Taking $\varphi = f$ and $v_+ = u_+$ in (5.41), we obtain

$$p^2 \|B_+^{1/2} u_+\|_{0;S^+}^2 + a_+(u_+, u_+) = (\mathcal{T}_p^+ f, f)_{0;\partial S}. \qquad (5.44)$$

Separating the real and imaginary parts in (5.44), we arrive at

$$|p|^2 \|B_+^{1/2} u_+\|_{0;S^+}^2 + a_+(u_+, u_+)$$
$$= \sigma^{-1} \operatorname{Re} \{\bar{p}(\mathcal{T}_p^+ f, f)_{0;\partial S}\}, \quad p = \sigma + i\tau,$$

and then, finally, at

$$\|u_+\|_{1,p;S^+}^2 \leq c|p| |(\mathcal{T}_p^+ f, f)_{0;\partial S}| \quad \forall p \in \bar{\mathbb{C}}_\kappa.$$

The rest of the argument is a verbatim repeat of the proof of Lemma 3.1. □

We define the single-layer potential $V_p \hat{\alpha}$ of density $\hat{\alpha} \in H_{-1/2,p}(\partial S)$, $\hat{\alpha} = \{\hat{\alpha}_1, \hat{\alpha}_2\}$, by setting

$$(V_p \hat{\alpha})(x, p) = (V_{p,1} \hat{\alpha}_1)(x, p) + (V_{p,2} \hat{\alpha}_2)(x, p), \quad x \in \mathbb{R}^2,$$

where

$$(V_{p,\nu} \hat{\alpha}_\nu)(x, p) = \int\limits_{\partial S_\nu} \hat{D}(x - y, p) \hat{\alpha}_\nu(y, p) \, ds_y, \quad \nu = 1, 2 \text{ (not summed)},$$

and define a boundary operator $V_{p,0}$ by

$$V_{p,0} \hat{\alpha} = \{(V_{p,0} \hat{\alpha})_1, (V_{p,0} \hat{\alpha})_2\},$$
$$(V_{p,0} \hat{\alpha})_\nu(x, p) = (V_p \hat{\alpha})(x, p), \quad x \in \partial S_\nu, \quad \nu = 1, 2.$$

The second equality above can also be written as

$$(V_{p,0} \hat{\alpha})_\nu = \gamma_\nu^+ \pi^+ V_p \hat{\alpha}.$$

5.5 Lemma. *For any $p \in \mathbb{C}_0$, the operator $V_{p,0}$ is a homeomorphism from $H_{-1/2,p}(\partial S)$ to $H_{1/2,p}(\partial S)$, and for any $\hat{\alpha} \in H_{-1/2,p}(\partial S)$, $p \in \bar{\mathbb{C}}_\kappa$, $\kappa > 0$,*

$$\|V_{p,0} \hat{\alpha}\|_{1/2,p;\partial S} \leq c|p| \|\hat{\alpha}\|_{-1/2,p;\partial S},$$
$$\|\hat{\alpha}\|_{-1/2,p;\partial S} \leq c|p| \|V_{p,0} \hat{\alpha}\|_{1/2,p;\partial S}. \qquad (5.45)$$

Proof. We remark that, as follows from the properties of single-layer potentials, for any $\hat{\alpha} \in H_{-1/2,p}(\partial S)$ we have the jump formula

$$(\mathcal{T}_p^+ - \mathcal{T}_p^-) V_{p,0} \hat{\alpha} = \hat{\alpha}.$$

The argument now continues as in the proof of Lemma 4.1. □

Let π_i^- and π_e^- be the operators of restriction from $\mathbb{R}^2$ (or $\mathbb{R}^2 \setminus \partial S$) to S_i^- and S_e^-, respectively. We define the double-layer potential $W_p \hat{\beta}$ of density $\hat{\beta} \in H_{1/2,p}(\partial S)$, $\hat{\beta} = \{\hat{\beta}_1, \hat{\beta}_2\}$, by

$$(W_p \hat{\beta})(x, p) = (W_{p,1} \hat{\beta}_1)(x, p) + (W_{p,2} \hat{\beta}_2)(x, p), \quad x \in \mathbb{R}^2 \setminus \partial S,$$

where

$$(W_{p,\nu} \hat{\beta}_\nu)(x, p) = \int\limits_{\partial S_\nu} \hat{P}(x - y, p) \hat{\beta}_\nu(y, p) \, ds_y, \quad \nu = 1, 2 \text{ (not summed)},$$

and the boundary operators $W_p^\pm$ of its limiting values on ∂S by

$$W_p^\pm \hat{\beta} = \gamma^\pm (W_p \hat{\beta}),$$

where, for a function u defined on $\mathbb{R}^2$ (or on $\mathbb{R}^2 \setminus \partial S$),

$$\gamma^+ u = \{\gamma_1^+ \pi^+ u, \gamma_2^+ \pi^+ u\}, \quad \gamma^- u = \{\gamma_1^- \pi_i^- u, \gamma_2^- \pi_e^- u\}.$$

5.6 Lemma. *For any $p \in \mathbb{C}_0$, the operators $W_p^\pm$ are homeomorphisms from $H_{1/2,p}(\partial S)$ to $H_{1/2,p}(\partial S)$, and for any $\hat{\beta} \in H_{1/2,p}(\partial S)$, $p \in \bar{\mathbb{C}}_\kappa$, $\kappa > 0$,*

$$\|W_p^\pm \hat{\beta}\|_{1/2,p;\partial S} \leq c|p|^2 \|\hat{\beta}\|_{1/2,p;\partial S},$$

$$\|\hat{\beta}\|_{1/2,p;\partial S} \leq c|p|^2 \|W_p^\pm \hat{\beta}\|_{1/2,p;\partial S}.$$

Proof. The assertion follows from the equalities

$$(W_p \hat{\beta})(x, p) = \begin{cases} (V_p \mathcal{T}_p^- \hat{\beta})(x, p), & x \in S^+, \\ (V_p \mathcal{T}_p^+ \hat{\beta})(x, p), & x \in S_i^- \cup S_e^-, \end{cases}$$

which are proved as in Lemmas 4.4, 5.4, and 5.5.   □

We now define an operator N_p by setting

$$N_p = \mathcal{T}_p^+ W_p^+ = \mathcal{T}_p^- W_p^-.$$

5.7 Lemma. *For every $p \in \mathbb{C}_0$, the operator N_p is a homeomorphism from $H_{1/2,p}(\partial S)$ to $H_{-1/2,p}(\partial S)$, and for any $\hat{\beta} \in H_{1/2,p}(\partial S)$, $p \in \bar{\mathbb{C}}_\kappa$, $\kappa > 0$,*

$$\|N_p \hat{\beta}\|_{-1/2,p;\partial S} \leq c|p|^3 \|\hat{\beta}\|_{1/2,p;\partial S},$$

$$\|\hat{\beta}\|_{1/2,p;\partial S} \leq c|p| \|N_p \hat{\beta}\|_{-1/2,p;\partial S}.$$

(5.46)

This assertion is proved in exactly the same way as Theorem 4.8.

Let

$$f = \{f_1, f_2\} \in H_{1/2,p}(\partial S), \quad g = \{g_1, g_2\} = T_p^+ f \in H_{-1/2,p}(\partial S).$$

We define a pair of boundary operators χ_1 and χ_2 by

$$\chi_1 f = \{g_1, f_2\} \in H_{-1/2,p}(\partial S_1) \times H_{1/2,p}(\partial S_2),$$

$$\chi_2 g = \{g_1, f_2\} \in H_{-1/2,p}(\partial S_1) \times H_{1/2,p}(\partial S_2).$$

5.8 Lemma. *For any $p \in \mathbb{C}_0$, the operator χ_1 is a homeomorphism from $H_{1/2,p}(\partial S)$ to $H_{-1/2,p}(\partial S_1) \times H_{1/2,p}(\partial S_2)$, while χ_2 is a homeomorphism from $H_{-1/2,p}(\partial S)$ to $H_{-1/2,p}(\partial S_1) \times H_{1/2,p}(\partial S_2)$, and for any $p \in \mathbb{C}_\kappa$, $\kappa > 0$,*

$$\|\chi_1 f\|_{H_{-1,2,p}(\partial S_1) \times H_{1/2,p}(\partial S_2)} \le c|p| \|f\|_{1/2,p;\partial S}, \tag{5.47}$$

$$\|f\|_{1/2,p;\partial S} \le c|p| \|\chi_1 f\|_{H_{-1/2,p}(\partial S_1) \times H_{1/2,p}(\partial S_2)}, \tag{5.48}$$

$$\|\chi_2 g\|_{H_{-1/2,p}(\partial S_1) \times H_{1/2,p}(\partial S_2)} \le c|p| \|g\|_{-1/2,p;\partial S}, \tag{5.49}$$

$$\|g\|_{-1/2,p;\partial S} \le c|p| \|\chi_2 g\|_{H_{-1/2,p}(\partial S_1) \times H_{1/2,p}(\partial S_2)}. \tag{5.50}$$

Proof. The continuity of χ_ν, $\nu = 1, 2$, follows from Lemma 5.4. Estimates (5.47) and (5.49) follow from (5.42) and (5.43), respectively. Let $\{g_1, f_2\} \in H_{-1/2,p}(\partial S_1) \times H_{1/2,p}(\partial S_2)$, and let $u \in H_{1,p}(S^+)$ be the (unique) solution of (5.32) with boundary data $\{g_1, f_2\}$. It is obvious that $f = \{\gamma_1^+ u, f_2\}$ satisfies $\chi_1 f = \{g_1, f_2\}$. Since (5.48) follows from (5.40), the statement for χ_1 is proved.

The function $g = T_p^+ f$ satisfies $\chi_2 g = \{g_1, f_2\}$, so χ_2 is a homeomorphism from $H_{-1/2,p}(\partial S)$ to $H_{-1/2,p}(\partial S_1) \times H_{1/2,p}(\partial S_2)$. To complete the proof, we need to show that (5.50) holds. Again, let u be the solution of (5.32) with boundary data $\{g_1, f_2\}$, $f = \{\gamma_1^+ u, f_2\}$, and $g = T_p^+ f$. By the definition of the operators T_p^+,

$$|(T_p^+ f, \varphi)_{0;\partial S}| \le c\|u\|_{1,p;S^+} \|\varphi\|_{1/2,p;\partial S};$$

hence,

$$\|g\|_{-1/2,p;\partial S} \le c\|u\|_{1,p;S^+},$$

and (5.50) follows from (5.40).   □

We introduce the time-dependent (or retarded) single-layer and double-layer potentials $V\alpha$ and $W\beta$ and their corresponding boundary operators V_0, $W^\pm$, $T^\pm$, and N in the spaces of originals in the usual way, and represent the solution u of problem (DMCD) in the form

$$u(X) = (V\alpha)(X), \quad X \in G^+, \tag{5.51}$$

or

$$u(X) = (W\beta)(X), \quad X \in G^+. \tag{5.52}$$

Representations (5.51) and (5.52) yield, respectively, the systems of nonstationary boundary integral equations

$$V_0 \alpha = f \tag{5.53}$$

and

$$W^+ \beta = f. \tag{5.54}$$

5.9 Theorem. *For any $f \in H_{1/2,k,\kappa}(\Gamma)$, $\kappa > 0$, $k \in \mathbb{R}$, systems (5.53) and (5.54) have unique solutions $\alpha \in H_{-1/2,k-1,\kappa}(\Gamma)$ and $\beta \in H_{1/2,k-2,\kappa}(\Gamma)$, respectively, in which case u defined by (5.51) or (5.52) belongs to $H_{1,k-1,\kappa}(G^+)$. If $k \geq 1$, then u is the weak solution of problem* (DMCD).

This assertion is proved by repeating the arguments in Theorem 4.9 and using Lemmas 5.5 and 5.6.

If we seek the solution u of (DMCN) in the form (5.51) or (5.52), then we arrive at the systems of boundary equations

$$\mathcal{T}^+ V_0 \alpha = g \tag{5.55}$$

and

$$N\beta = g. \tag{5.56}$$

5.10 Theorem. *For any $g \in H_{-1/2,k,\kappa}(\Gamma)$, $\kappa > 0$, $k \in \mathbb{R}$, systems (5.55) and (5.56) have unique solutions $\alpha \in H_{-1/2,k-2,\kappa}(\Gamma)$ and $\beta \in H_{1/2,k-1\kappa}(\Gamma)$, in which case u defined by (5.51) or (5.52) belongs to $H_{1,k-1,\kappa}(G^+)$. If $k \geq 1$, then u is the weak solution of problem* (DMCN).

This assertion is proved just like Theorem 4.10, use being made of Lemmas 5.4, 5.5, and 5.7.

Finally, we seek the solution u of (DMCM) in the form (5.51) or (5.52) and arrive at the systems of boundary equations

$$(\mathcal{T}^+ V_0 \alpha)_1 = g_1, \quad (V_0 \alpha)_2 = f_2 \tag{5.57}$$

and

$$(N\beta)_1 = g_1, \quad (W^+ \beta)_2 = f_2, \tag{5.58}$$

respectively.

5.11 Theorem. *For any $\{g_1, f_2\} \in H_{-1/2,k,\kappa}(\Gamma) \times H_{1/2,k,\kappa}(\Gamma)$, $\kappa > 0$, $k \in \mathbb{R}$, systems (5.57) and (5.58) have unique solutions $\alpha \in H_{-1/2,k-2,\kappa}(\Gamma)$ and $\beta \in H_{1/2,k-2,\kappa}(\Gamma)$, in which case u defined by (5.51) or (5.52) belongs to $H_{1,k-1,\kappa}(G^+)$. If $k \geq 1$, then u is the weak solution of problem* (DMCM).

Proof. In terms of Laplace transforms, (5.57) and (5.58) become

$$(\mathcal{T}_p^+ V_{p,0}\hat{\alpha})_1 = \hat{g}_1, \quad (V_{p,0}\hat{\alpha})_2 = \hat{f}_2$$

and

$$(N_p\hat{\beta})_1 = \hat{g}_1, \quad (W_p^+\hat{\beta})_2 = \hat{f}_2,$$

or, equivalently,

$$\chi_1 V_{p,0}\hat{\alpha} = \{\hat{g}_1, \hat{f}_2\} \tag{5.59}$$

and

$$\chi_2 N_p\hat{\beta} = \{\hat{g}_1, \hat{f}_2\}. \tag{5.60}$$

By Lemmas 5.5, 5.7, and 5.8, systems (5.59) and (5.60) have unique solutions

$$\hat{\alpha} = V_{p,0}^{-1}\chi_1^{-1}\{\hat{g}_1, \hat{f}_2\} \in H_{-1/2,p}(\partial S) \tag{5.61}$$

and

$$\hat{\beta} = N_p^{-1}\chi_2^{-1}\{\hat{g}_1, \hat{f}_2\} \in H_{1/2,p}(\partial S). \tag{5.62}$$

In turn, by (5.61), (5.62), (5.45), (5.46), (5.48), and (5.50),

$$\|\hat{\alpha}\|_{-1/2,p;\partial S} \leq c|p|^2\{\|\hat{g}_1\|_{-1/2,p;\partial S} + \|\hat{f}_2\|_{1/2,p;\partial S}\}, \tag{5.63}$$

$$\|\hat{\beta}\|_{1/2,p;\partial S} \leq c|p|^2\{\|\hat{g}_1\|_{-1/2,p;\partial S} + \|\hat{f}_2\|_{1/2,p;\partial S}\}. \tag{5.64}$$

The rest of the proof now follows from (5.63), (5.64), and Theorem 5.3. $\quad\square$

5.3 Finite Plate with an Inclusion

Keeping the notation used in §5.2, below we consider an initial-boundary value problem for a piecewise homogeneous finite plate occupying the domain $S = \bar{S}_i^- \cup S^+$. To simplify the symbols, we write S^- instead of S_i^-. We assume that the plates with middle sections S^+ and S^- have Lamé constants λ_+, μ_+ and λ_-, μ_-, densities ρ_+, ρ_-, and thickness parameters h_+, h_-. If u is a function defined on S, then we write $u = \{u_+, u_-\}$, where $u_\pm$ are the restrictions of u from S to $S^\pm$. The same notation is used if u is defined in the domain G.

Let (DTD) be the initial-boundary value problem with transmission conditions across ∂S_1 and Dirichlet condition on ∂S_2 which consists in finding

$$u_+ \in C^2(G^+) \cap C^1(\bar{G}^+), \quad u_- \in C^2(G^-) \cap C^1(\bar{G}^-)$$

such that

$$B_+(\partial_t^2 u_+)(X) + (A_+ u_+)(X) = 0, \quad X \in G^+,$$
$$B_-(\partial_t^2 u_-)(X) + (A_- u_-)(X) = 0, \quad X \in G^-,$$
$$u_+(x, 0+) = (\partial_t u_+)(x, 0+) = 0, \quad x \in S^+,$$
$$u_-(x, 0+) = (\partial_t u_-)(x, 0+) = 0, \quad x \in S^-,$$
$$u_+^+(X) - u_-^-(X) = f_1(X), \quad X \in \Gamma_1,$$
$$(T_+ u_+)^+(X) - (T_- u_-)^-(X) = g_1(X), \quad X \in \Gamma_1,$$
$$u_+^+(X) = f_2(X), \quad X \in \Gamma_2,$$

where f_1, f_2, and g_1 are prescribed.

Proceeding as in §5.1, we readily see that in the variational version of (DTD) we seek $u = \{u_+, u_-\}$, $u_\pm \in H_{1,0,\kappa}(G^\pm)$, satisfying

$$\int_0^\infty \big[a_+(u_+, v_+) + a_-(u_-, v_-)$$
$$- (B_+^{1/2} \partial_t u_+, B_+^{1/2} \partial_t v_+)_{0;S^+} - (B_-^{1/2} \partial_t u_-, B_-^{1/2} \partial_t v_-)_{0;S^-} \big] \, dt$$
$$= \int_0^\infty (g_1, v_0)_{0;\partial S_1} \, dt \quad \forall v \in C_0^\infty(S \times [0, \infty)), \tag{5.65}$$
$$\gamma_1^+ u_+ - \gamma_1^- u_- = f_1, \quad \gamma_2^+ u_+ = f_2,$$

where the trace operators $\gamma_1^\pm$ and γ_2^+ have the obvious meaning, $v_0 = (v_+^+)_1 = v_-^-$, and $f = \{f_1, f_2\} \in H_{1/2,k,\kappa}(\Gamma)$ and $g_1 \in H_{-1/2,k,\kappa}(\Gamma_1)$ are prescribed.

5.12 Theorem. *For any $f \in H_{1/2,1,\kappa}(\Gamma)$ and $g_1 \in H_{-1/2,1,\kappa}(\Gamma_1)$, $\kappa > 0$, problem (5.65) has a unique solution $u = \{u_+, u_-\}$, where $u_\pm \in H_{1,0,\kappa}(G^\pm)$. If $f \in H_{1/2,k,\kappa}(\Gamma)$ and $g_1 \in H_{-1/2,k,\kappa}(\Gamma_1)$, $k \in \mathbb{R}$, then $u_\pm \in H_{1,k-1,\kappa}(G^\pm)$ and*

$$\|u_+\|_{1,k-1,\kappa;G^+} + \|u_-\|_{1,k-1,\kappa;G^-} \le c(\|f\|_{1/2,k,\kappa;\Gamma} + \|g_1\|_{-1/2,k,\kappa;\Gamma_1}).$$

The assertion is proved exactly as Theorem 5.1, with $H_{1,p}(\mathbb{R}^2)$ replaced by $\overset{\circ}{H}_{1,p}(S)$ and (5.12) written for this case, where it takes the form

$$\|\hat{u}_+\|_{1,p;S^+} + \|\hat{u}_-\|_{1,p;S^-} \le c|p|\{\|\hat{f}\|_{1/2,p;\partial S} + \|\hat{g}_1\|_{-1/2,p;\partial S_1}\}. \tag{5.66}$$

We now represent the solution of (5.65) in terms of single-layer potentials. Let $V_- \alpha_-$ be the time-dependent (retarded) single-layer potential defined by

$$(V_- \alpha_-)(X) = \int_{\partial S_1} \int_0^\infty D_-(x - y, t - \tau)\alpha_-(y, \tau) \, ds_y \, d\tau, \quad X \in G^-,$$

where α_- is an unknown density defined on Γ_1 and D_- is the matrix of fundamental solutions of the equation of motion for the plate corresponding to S^-. We also consider the single-layer potential $V_+\alpha_+$ defined by

$$(V_+\alpha_+)(X) = \int_{\partial S} \int_0^\infty D_+(x - y, t - \tau)\alpha_+(y, \tau)\, ds_y\, d\tau, \quad X \in G^+,$$

where

$$\alpha_+ = \{\alpha_{+,1}, \alpha_{+,2}\},$$

$\alpha_{+,1}$ and $\alpha_{+,2}$ are unknown densities defined on Γ_1 and Γ_2, respectively, and D_+ is the matrix of fundamental solutions for the plate with middle plane section S^+. The properties of $V_+\alpha_+$ were studied in §5.2. The boundary operators generated by these potentials are denoted by $V_{+,0}$ and $V_{-,0}$, where

$$V_{+,0}\alpha_+ = \{(V_{+,0}\alpha_+)_1, (V_{+,0}\alpha_+)_2\}.$$

Analogous notation W_+^+ and W_-^- is used for the double-layer potentials $W_+\beta_+$ and $W_-\beta_-$, where

$$W_+^+\beta_+ = \{(W_+^+\beta_+)_1, (W_+^+\beta_+)_2\}$$

are the boundary operators generated by the limiting values of $W_+\beta_+$ and $W_-\beta_-$ on Γ. Finally, we use the symbols $\mathcal{T}_+^+$ and $\mathcal{T}_+^-$ for the corresponding Poincaré–Steklov operators, and N_+ and N_- for the operators of the moment-force boundary values generated by the double-layer potentials. In the spaces of Laplace transforms, all the above symbols acquire an extra subscript p.

We seek the solution $u = \{u_+, u_-\}$ of (5.65) in the form

$$\begin{aligned}
u_+(X) &= (V_+\alpha_+)(X), \quad X \in G^+, \\
u_-(X) &= (V_-\alpha_-)(X), \quad X \in G^-.
\end{aligned} \tag{5.67}$$

Representation (5.67) yields the system of boundary integral equations

$$\begin{aligned}
(V_{+,0}\alpha_+)_1 - V_{-,0}\alpha_- &= f_1, \\
(\mathcal{T}_+^+ V_{+,0}\alpha_+)_1 - \mathcal{T}_-^- V_{-,0}\alpha_- &= g_1, \\
(V_{+,0}\alpha_+)_2 &= f_2.
\end{aligned} \tag{5.68}$$

5.13 Theorem. *For any $f \in H_{1/2,k,\kappa}(\Gamma)$ and $g_1 \in H_{-1/2,k,\kappa}(\Gamma_1)$, where $\kappa > 0$ and $k \in \mathbb{R}$, system (5.68) has a unique solution $\alpha_+ \in H_{-1/2,k-2,\kappa}(\Gamma)$, $\alpha_- \in H_{-1/2,k-2,\kappa}(\Gamma_1)$, in which case u defined by (5.67) belongs to the space $H_{1,k-1,\kappa}(G^+) \times H_{1,k-1,\kappa}(G^-)$. If $k \geq 1$, then u is the weak solution of problem (DTD).*

Proof. In terms of Laplace transforms, (5.68) takes the form

$$(V_{p,+,0}\hat{\alpha}_+)_1 - V_{p,-,0}\hat{\alpha}_- = \hat{f}_1,$$
$$(T^+_{p,+}V_{p,+,0}\hat{\alpha}_+)_1 - T^-_{p,-}V_{p,-,0}\hat{\alpha}_- = \hat{g}_1, \qquad (5.69)$$
$$(V_{p,+,0}\hat{\alpha}_+)_2 = \hat{f}_2.$$

Next, (5.65) turns into the problem of finding $\hat{u} = \{\hat{u}_+, \hat{u}_-\}$, where $\hat{u}_\pm \in H_{1,p}(S^\pm)$, such that

$$
\begin{aligned}
p^2(B^{1/2}_+\hat{u}_+, B^{1/2}_+v_+)_{0;S^+} &+ p^2(B^{1/2}_-\hat{u}_-, B^{1/2}_-v_-)_{0;S^-} \\
&+ a_+(\hat{u}_+, v_+) + a_-(\hat{u}_-, v_-) \\
&= (\hat{g}_1, v_0)_{0;\partial S_1} \quad \forall v \in \mathring{H}_{1,p}(S), \\
\gamma^+_1\hat{u}_+ - \gamma^-_1\hat{u}_- &= \hat{f}_1, \quad \gamma^+_2\hat{u}_+ = \hat{f}_2.
\end{aligned}
\qquad (5.70)
$$

We recall that the solution $\hat{u}$ of (5.70) satisfies (5.66). We write $\gamma^+_1\hat{u}_+ = \hat{f}_+$, $\gamma^-_1\hat{u}_- = \hat{f}_-$ and solve the systems of equations

$$
\begin{aligned}
V_{p,+,0}\hat{\alpha}_+ &= \{\hat{f}_+, \hat{f}_2\}, \\
V_{p,-,0}\hat{\alpha}_- &= \hat{f}_-.
\end{aligned}
\qquad (5.71)
$$

From the properties of $V_{p,+,0}$ and $V_{p,-,0}$ it follows that (5.71) has a unique solution $\{\hat{\alpha}_+, \hat{\alpha}_-\}$ for any $p \in \mathbb{C}_\kappa$, and that

$$\|\hat{\alpha}_+\|_{-1/2,p;\partial S} \le c|p|\|\{\hat{f}_+, \hat{f}_2\}\|_{1/2,p;\partial S},$$

$$\|\hat{\alpha}_-\|_{-1/2,p;\partial S_1} \le c|p|\|\hat{f}_-\|_{1/2,p;\partial S_1}.$$

The trace theorem and (5.66) imply that

$$\|\hat{\alpha}_+\|_{-1/2,p;\partial S} \le c|p|\|\hat{u}_+\|_{1,p;S^+}$$
$$\le c|p|^2(\|\hat{f}\|_{1/2,p;\partial S} + \|\hat{g}_1\|_{-1/2,p;\partial S_1}), \qquad (5.72)$$
$$\|\hat{\alpha}_-\|_{-1/2,p;\partial S_1} \le c|p|\|\hat{u}_-\|_{1,p;S^-}$$
$$\le c|p|^2(\|\hat{f}\|_{1/2,p;\partial S} + \|\hat{g}_1\|_{-1/2,p;\partial S_1}). \qquad (5.73)$$

Remarking that $\{\hat{\alpha}_+, \hat{\alpha}_-\}$ satisfies (5.69) and taking (5.72) and (5.73) into account, we complete the proof by means of the standard scheme used in the case of the analogous assertions in the preceding theorems. $\square$

If we now seek the solution of (5.65) in the form

$$
\begin{aligned}
u_+(X) &= (W_+\beta_+)(X), \quad X \in G^+, \\
u_-(X) &= (W_-\beta_-)(X), \quad X \in G^-,
\end{aligned}
\qquad (5.74)
$$

then we arrive at the system of boundary equations

$$(W_+^+\beta_+)_1 - W_-^-\beta_- = f_1,$$
$$(N_+\beta_+)_1 - N_-\beta_- = g_1, \tag{5.75}$$
$$(W_+^+\beta_+)_2 = f_2.$$

5.14 Theorem. *For any* $f \in H_{1/2,k,\kappa}(\Gamma)$ *and* $g_1 \in H_{-1/2,k,\kappa}(\Gamma_1)$, *where* $\kappa > 0$ *and* $k \in \mathbb{R}$, *system* (5.75) *has a unique solution* $\beta_+ \in H_{1/2,k-2,\kappa}(\Gamma)$, $\beta_- \in H_{1/2,k-2,\kappa}(\Gamma_1)$, *in which case* u *defined by* (5.74) *belongs to the space* $H_{1,k-1,\kappa}(G^+) \times H_{1,k-1,\kappa}(G^-)$. *If* $k \geq 1$, *then* u *is the weak solution of problem* (DTD).

Proof. In terms of Laplace transforms, (5.75) becomes

$$(W_{p,+}^+\hat{\beta}_+)_1 - W_{p,-}^-\hat{\beta}_- = \hat{f}_1,$$
$$(N_{p,+}\hat{\beta}_+)_1 - N_{p,-}\hat{\beta}_- = \hat{g}_1, \tag{5.76}$$
$$(W_{p,+}^+\hat{\beta}_+)_2 = \hat{f}_2.$$

Once more, let $\hat{u}$ be the solution of (5.70), and let

$$\hat{g}_+ = \mathcal{T}_{p,+}^+\{\hat{f}_+, \hat{f}_2\}, \quad \hat{g}_- = \mathcal{T}_{p,-}^-\hat{f}_-,$$

where $\hat{f}_+ = \gamma_1^+\hat{u}_+$ and $\hat{f}_- = \gamma_1^-\hat{u}_-$. It is obvious that $\{\hat{\beta}_+, \hat{\beta}_-\}$, where

$$\hat{\beta}_\pm = N_{p,\pm}^{-1}\hat{g}_\pm, \tag{5.77}$$

is a solution of (5.76). The properties of $N_{p,\pm}^{-1}$ and (5.77) imply that for any $p \in \mathbb{C}_\kappa$,

$$\|\hat{\beta}_+\|_{1/2,p;\partial S} \leq c|p|\|\hat{g}_+\|_{-1/2,p;\partial S},$$
$$\|\hat{\beta}_-\|_{1/2,p;\partial S_1} \leq c|p|\|\hat{g}_-\|_{-1/2,p;\partial S_1}.$$

Since

$$\|\hat{g}_+\|_{-1/2,p;\partial S} \leq c\|\hat{u}_+\|_{1,p;S^+},$$
$$\|\hat{g}_-\|_{-1/2,p;\partial S_1} \leq c\|\hat{u}_-\|_{1,p;S^-},$$

from (5.66) it follows that

$$\|\hat{\beta}_+\|_{1/2,p;\partial S} + \|\hat{\beta}_-\|_{1/2,p;\partial S_1} \leq c|p|^2(\|\hat{f}\|_{1/2,p;\partial S} + \|\hat{g}_1\|_{-1/2,p;\partial S_1}).$$

The rest of the proof is now completed in the usual way. $\square$

Plate Weakened by a Crack

6.1 Formulation and Solvability of the Problems

Let ∂S_0 be an open arc of a simple closed C^2-curve ∂S in $\mathbb{R}^2$, and let

$$\partial S_1 = \partial S \setminus \overline{\partial S_0}, \quad S = \mathbb{R}^2 \setminus \overline{\partial S_0}.$$

We denote by π_i, $i = 0, 1$, the operators of restriction from ∂S to ∂S_i. As usual, $S^\pm$ are the interior and exterior domains bounded by ∂S.

We make the notation

$$G = S \times (0, \infty), \quad G^\pm = S^\pm \times (0, \infty),$$

$$\Gamma = \partial S \times (0, \infty), \quad \Gamma_i = \partial S_i \times (0, \infty), \quad i = 0, 1.$$

The operators of restriction from S (or $\mathbb{R}^2$) to $S^\pm$, or from G (or $\mathbb{R}^2 \times (0, \infty)$) to $G^\pm$, are denoted by $\pi^\pm$.

If u is a function defined on S, then we write

$$u_\pm = \pi^\pm u, \quad u = (u_+, u_-).$$

Let $\gamma_i^\pm = \pi_i \gamma^\pm$, $i = 0, 1$, be the operators of trace on ∂S_i (or Γ_i) for functions defined on $S^\pm$ (or $G^\pm$).

$H_{1,p}(S)$, $p \in \mathbb{C}$, is the space of all $u(x, p) = (u_+(x, p), u_-(x, p))$ such that $u_\pm \in H_{1,p}(S^\pm)$ and $\gamma_1^+ u_+ = \gamma_1^- u_-$. This space is equipped with the norm

$$\|u\|_{1,p;S} = \|u_+\|_{1,p;S^+} + \|u_-\|_{1,p;S^-}.$$

$\mathring{H}_{-1,p}(S)$ is the dual of $H_{1,p}(S)$ with respect to the duality generated by the inner product $(\cdot, \cdot)_{0;S}$ in $L^2(S)$.

$\mathring{H}_{1,p}(S)$ is the subspace of $H_{1,p}(\mathbb{R}^2)$ that consists of all u such that $\gamma_0^+ u_+ = \gamma_0^- u_- = 0$.

$H_{-1,p}(S)$ is the dual of $\mathring{H}_{1,p}(S)$ with respect to the duality generated by $(\cdot,\cdot)_{0;S}$.

$\mathring{H}_{1/2,p}(\partial S_0)$ is the subspace of all functions $f \in H_{1/2,p}(\partial S)$ such that $\operatorname{supp} f \in \overline{\partial S_0}$.

$H_{-1/2,p}(\partial S_0)$ is the dual of $\mathring{H}_{1/2,p}(\partial S_0)$ with respect to the duality generated by the inner product $(\cdot,\cdot)_{0;\partial S_0}$ in $L^2(\partial S_0)$.

$H_{1/2,p}(\partial S_0)$ is the space of the restrictions φ from ∂S to ∂S_0 of all $f \in H_{1/2,p}(\partial S)$, equipped with the norm

$$\|\varphi\|_{1/2,p;\partial S_0} = \inf_{f \in H_{1/2,p}(\partial S):\, \pi_0 f = \varphi} \|f\|_{1/2,p;\partial S}.$$

$\mathring{H}_{-1/2,p}(\partial S_0)$ is the dual of $H_{1/2,p}(\partial S_0)$ with respect to the duality generated by $(\cdot,\cdot)_{0;\partial S_0}$.

$H_{\pm 1,k,\kappa}(G)$, $H_{\pm 1/2,k,\kappa}(\Gamma_0)$, and $\mathring{H}_{\pm 1/2,k,\kappa}(\Gamma_0)$, $k \in \mathbb{R}$, $\kappa > 0$, and the norms on these spaces, are now introduced in the usual way. We remark that the trace operators $\gamma_0^{\pm}$ map $H_{1,k,\kappa}(G^{\pm})$ continuously onto $H_{1/2,k,\kappa}(\Gamma_0)$. We also remark that if $u = (u_+, u_-) \in H_{1,k,\kappa}(G)$, then $\gamma_0^+ u - \gamma_0^- u \in \mathring{H}_{1/2,k,\kappa}(\Gamma_0)$, where

$$\gamma_0^{\pm} u = \gamma_0^{\pm} u_{\pm}.$$

In what follows, we denote by $a_S(u,v)$ the energy bilinear form constructed for the plate with middle plane domain S.

In the variational version of the initial-boundary value problem (DKD), we seek $u \in H_{1,0,\kappa}(G)$ satisfying

$$\int_0^\infty \left[a_S(u,v) - (B^{1/2}\partial_t u, B^{1/2}\partial_t v)_{0;S}\right] dt = 0$$

$$\forall v \in C_0^\infty(S \times [0,\infty)), \qquad (6.1)$$

$$\gamma_0^+ u = f_+, \quad \gamma_0^- u = f_-,$$

where f_+ and f_- are prescribed on Γ_0.

6.1 Theorem. *For any given f_+, $f_- \in H_{1/2,1,\kappa}(\Gamma_0)$, $\kappa > 0$, such that $\delta f = f_+ - f_- \in \mathring{H}_{1/2,1,\kappa}(\Gamma_0)$, problem (6.1) has a unique solution $u \in H_{1,0,\kappa}(G)$. If f_+, $f_- \in H_{1/2,k,\kappa}(\Gamma_0)$ and $\delta f \in \mathring{H}_{1/2,k,\kappa}(\Gamma_0)$, $k \in \mathbb{R}$, then $u \in H_{1,k-1,\kappa}(G)$ and*

$$\|u\|_{1,k-1,\kappa;G} \le c\big(\|f_+\|_{1/2,k,\kappa;\Gamma_0} + \|\delta f\|_{1/2,k,\kappa;\Gamma}\big).$$

Proof. Going over to Laplace transforms in (6.1), we arrive at a problem that consists in finding $u \in H_{1,p}(S)$ such that

$$p^2(B^{1/2}u, B^{1/2}v)_{0;S} + a_S(u,v) = 0 \quad \forall v \in \mathring{H}_{1,p}(S),$$

$$\gamma_0^+ u = f_+, \quad \gamma_0^- u = f_-. \qquad (6.2)$$

First, we study an auxiliary problem, in which we want to find $w \in \mathring{H}_{1,p}(S)$ such that

$$p^2(B^{1/2}w, B^{1/2}v)_{0;S} + a_S(w, v) = (q, v)_{0;S} \quad \forall v \in \mathring{H}_{1,p}(S), \tag{6.3}$$

where $q \in H_{-1,p}(S)$. The unique solvability of (6.3) is proved by means of standard arguments. In Theorem 2.1 it was shown that the form $a_S(w, v)$ is symmetric, continuous, and coercive on $[\mathring{H}_{1,p}(S)]^2$, that, consequently, it defines a self-adjoint operator $\mathcal{A}$ on $\mathring{H}_{1,p}(S)$ through the equality

$$(\mathcal{A}u, v)_{0;S} = a_S(u, v) \quad \forall u, v \in \mathring{H}_{1,p}(S),$$

and that $\mathcal{A}$ is a homeomorphism from $\mathring{H}_{1,p}(S)$ to $H_{-1,p}(S)$. We now rewrite (6.3) as

$$\mathcal{A}w + p^2 Bw = q,$$

or

$$w_s + p^2 \mathcal{B} w_s = B^{1/2} \mathcal{A}^{-1} q, \tag{6.4}$$

where $w_s = B^{1/2}w$ and $\mathcal{B} = B^{1/2}\mathcal{A}^{-1}B^{1/2}$. Next, we verify that equation (6.4) in $\mathring{H}_{1,p}(S)$ is equivalent to itself in $L^2(S)$, and that $\mathcal{B}$ is a self-adjoint nonnegative operator on $L^2(S)$. This proves that (6.4)—hence, also (6.3)—have a unique solution $w \in \mathring{H}_{1,p}(S)$ for any $q \in H_{-1,p}(S)$. Setting $v = w$ in (6.3) and separating the real and imaginary parts, we obtain the estimate

$$\|w\|_{1,p} \leq c|p| \|q\|_{-1,p;S}. \tag{6.5}$$

Let l_0 be an extension operator from ∂S_0 to ∂S, which maps $H_{1/2,p}(\partial S_0)$ continuously to $H_{1/2,p}(\partial S)$; that is,

$$\|l_0 f_+\|_{1/2,p;\partial S} \leq c\|f_+\|_{1/2,p;\partial S_0} \quad \forall f_+ \in H_{1/2,p}(\partial S_0).$$

Let $F_+ = l_0 f_+$, and let F_- be an extension of f_- from ∂S_0 to ∂S such that $\pi_1 F_+ = \pi_1 F_-$. We take $u_0 = (l^+ F_+, l^- F_-) \in H_{1,p}(S)$ and seek a solution u of (6.2) in the form $u = u_0 + w$. Clearly, $w \in \mathring{H}_{1,p}(S)$ satisfies

$$p^2(B^{1/2}w, B^{1/2}v)_{0;S} + a_S(w, v)$$
$$= -p^2(B^{1/2}u_0, B^{1/2}v)_{0;S} - a_S(u_0, v) \quad \forall v \in \mathring{H}_{1,p}(S). \tag{6.6}$$

Since for any $v \in \mathring{H}_{1,p}(S)$,

$$|p^2(B^{1/2}u_0, B^{1/2}v)_{0;S} + a_S(u_0, v)| \leq c\|u_0\|_{1,p;S} \|v\|_{1,p}$$
$$\leq c(\|F_+\|_{1/2,p;\partial S} + \|F_-\|_{1/2,p;\partial S}) \|v\|_{1,p}$$
$$\leq c(\|f_+\|_{1/2,p;\partial S_0} + \|\delta f\|_{1/2,p;\partial S}) \|v\|_{1,p},$$

we can write the right-hand side in (6.6) as $(q, v)_{0;S}$, where $q \in H_{-1,p}(S)$ and

$$\|q\|_{-1,p;S} \leq c(\|f_+\|_{1/2,p;\partial S_0} + \|\delta f\|_{1/2,p;\partial S}).$$

Therefore, (6.2) has a unique solution $u \in H_{1,p}(S)$ and, as follows from (6.5),

$$\|u\|_{1,p;S} \leq c|p|(\|f_+\|_{1/2,p;\partial S_0} + \|\delta f\|_{1/2,p;\partial S}). \tag{6.7}$$

Using (6.7), we complete the proof in the standard way. $\square$

Let $C_0^\infty(\bar{G})$ be the space of all functions with compact support in $\bar{G}$ which belong to both $C_0^\infty(\bar{G}^+)$ and $C_0^\infty(\bar{G}^-)$ and are such that their limiting values, and the limiting values of all their derivatives, on Γ_1 from inside G^+ and G^- coincide.

In the variational version of problem (DKN), we seek $u \in H_{1,0,\kappa}(G)$ that satisfies

$$\int_0^\infty \left[a_S(u, v) - (B^{1/2}\partial_t u, B^{1/2}\partial_t v)_{0;S}\right] dt$$

$$= \int_0^\infty \left[(g_+, \gamma_0^+ v)_{0;\partial S_0} - (g_-, \gamma_0^- v)_{0;\partial S_0}\right] dt$$

$$\forall v \in C_0^\infty(\bar{G}), \tag{6.8}$$

where $g_\pm$ are prescribed on Γ_0.

6.2 Theorem. *For any given* $g_+, g_- \in H_{-1/2,1,\kappa}(\Gamma_0)$, $\kappa > 0$, *such that* $\delta g = g_+ - g_- \in \mathring{H}_{-1/2,1,\kappa}(\Gamma_0)$, *problem* (6.8) *has a unique solution* $u \in H_{1,0,\kappa}(G)$. *If* $g_+, g_- \in H_{-1/2,k,\kappa}(\Gamma_0)$ *and* $\delta g \in \mathring{H}_{-1/2,k,\kappa}(\Gamma_0)$, $k \in \mathbb{R}$, *then* $u \in H_{1,k-1,\kappa}(G)$ *and*

$$\|u\|_{1,k-1,\kappa;G} \leq c(\|\delta g\|_{-1/2,k,\kappa;\Gamma} + \|g_-\|_{-1/2,k,\kappa;\Gamma_0}).$$

Proof. In terms of Laplace transforms, (6.8) becomes

$$p^2(B^{1/2}u, B^{1/2}v)_{0;S} + a_S(u, v)$$

$$= (g_+, \gamma_0^+ v)_{0;\partial S_0} - (g_-, \gamma_0^- v)_{0;\partial S_0} \quad \forall v \in H_{1,p}(S), \tag{6.9}$$

or, which is the same,

$$p^2(B^{1/2}u, B^{1/2}v)_{0;S} + a_S(u, v)$$

$$= (\delta g, \gamma_0^+ v)_{0;\partial S_0} + (g_-, \gamma_0^+ v - \gamma_0^- v)_{0;\partial S_0} \quad \forall v \in H_{1,p}(S).$$

Since for any $v \in H_{1,p}(S)$,

$$\gamma_0^+ v - \gamma_0^- v \in \mathring{H}_{1/2,p}(\partial S_0),$$

it follows that

$$|(\delta g, \gamma_0^+ v)_{0;\partial S_0} + (g_-, \gamma_0^+ v - \gamma_0^- v)_{0;\partial S_0}|$$
$$\leq c(\|\delta g\|_{-1/2,p;\partial S}\|\gamma_0^+ v\|_{1/2,p;\partial S_0}$$
$$+ \|g_-\|_{-1/2,p;\partial S_0}\|\gamma_0^+ v - \gamma_0^- v\|_{1/2,p;\partial S})$$
$$\leq c(\|\delta g\|_{-1/2,p;\partial S} + \|g_-\|_{-1/2,p;\partial S_0})\|v\|_{1,p;S},$$

so we can write (6.9) as

$$p^2(B^{1/2}u, B^{1/2}v)_{0;S} + a_S(u,v) = (Q,v)_{0;S} \quad \forall v \in H_{1,p}(S), \tag{6.10}$$

where $Q \in \mathring{H}_{-1,p}(S)$ and

$$\|Q\|_{-1,p} \leq c(\|\delta g\|_{-1/2,p;\partial S} + \|g_-\|_{-1/2,p;\partial S_0}).$$

To prove that (6.10) is uniquely solvable, we consider the bilinear form

$$a_{\kappa,S}(u,v) = \tfrac{1}{2}\kappa^2(B^{1/2}u, B^{1/2}v)_{0;S} + a_S(u,v).$$

This form is symmetric, continuous, and coercive on $[H_{1,p}(S)]^2$; consequently, it defines a self-adjoint operator $\mathcal{A}_\kappa$, which is a homeomorphism from $H_{1,p}(S)$ to $\mathring{H}_{-1,p}(S)$. Repeating standard arguments, we show that (6.9) is uniquely solvable in $H_{1,p}(S)$ and that

$$\|u\|_{1,p;S} \leq c|p|(\|\delta g\|_{-1/2,p;\partial S} + \|g_-\|_{-1/2,p;\partial S_0}). \tag{6.11}$$

Taking (6.11) into account, we complete the proof by following the usual procedure. □

6.2 The Poincaré–Steklov Operator

In this section we introduce the Poincaré–Steklov operator $\mathcal{T}$ and study its properties. We start by defining this operator in Sobolev spaces with a parameter.

Let

$$\tilde{f} = \{f_+, \delta f\} \in H_{1/2,p}(\partial S_0) \times \mathring{H}_{1/2,p}(\partial S_0),$$
$$\tilde{g} = \{\delta g, g_-\} \in \mathring{H}_{-1/2,p}(\partial S_0) \times H_{-1/2,p}(\partial S_0).$$

We write

$$\langle \tilde{f}, \tilde{g} \rangle_{0;\partial S_0} = (f_+, \delta g)_{0;\partial S_0} + (\delta f, g_-)_{0;\partial S_0}.$$

The norms of $\tilde{f}$ and $\tilde{g}$ in $H_{1/2,p}(\partial S_0) \times \overset{\circ}{H}_{1/2,p}(\partial S_0)$ and $\overset{\circ}{H}_{-1/2,p}(\partial S_0) \times H_{-1/2,p}(\partial S_0)$, respectively, are defined by

$$\|\tilde{f}\|_{1/2,p;\partial S_0;1/2,p;\partial S} = \|f_+\|_{1/2,p;\partial S_0} + \|\delta f\|_{1/2,p;\partial S},$$

$$\|\tilde{g}\|_{-1/2,p;\partial S;-1/2,p;\partial S_0} = \|\delta g\|_{-1/2,p;\partial S} + \|g_-\|_{-1/2,p;\partial S_0}.$$

Let $\tilde{f}, \tilde{\varphi} \in H_{1/2,p}(\partial S_0) \times \overset{\circ}{H}_{1/2,p}(\partial S_0)$, $\tilde{\varphi} = \{\varphi_+, \delta\varphi\}$, let $u \in H_{1,p}(S)$ be the solution of (6.2) with $f_- = f_+ - \delta f$, and let $v \in H_{1,p}(S)$ be such that $\gamma_0^+ v = \varphi_+$ and $\gamma_0^+ v - \gamma_0^- v = \delta\varphi$. For every $p \in \mathbb{C}_0$ we define the Poincaré–Steklov operator $\mathcal{T}_p$ by means of the equality

$$\langle \mathcal{T}_p\tilde{f}, \tilde{g}\rangle_{0;\partial S_0} = p^2(B^{1/2}u, B^{1/2}v)_{0;S} + a_S(u,v). \tag{6.12}$$

It is easy to convince ourselves that (6.12) defines $\mathcal{T}_p$ correctly; that is, its definition does not depend on the choice of v.

6.3 Lemma. *For any $p \in \mathbb{C}_0$, the operator $\mathcal{T}_p$ is a homeomorphism from the space $H_{1/2,p}(\partial S_0) \times \overset{\circ}{H}_{1/2,p}(\partial S_0)$ to its dual $\overset{\circ}{H}_{-1/2,p}(\partial S_0) \times H_{-1/2,p}(\partial S_0)$, and for any $\tilde{f} \in H_{1/2,p}(\partial S_0) \times \overset{\circ}{H}_{1/2,p}(\partial S_0)$, $p \in \bar{\mathbb{C}}_\kappa$, $\kappa > 0$,*

$$\|\mathcal{T}_p\tilde{f}\|_{-1/2,p;\partial S;-1/2,p;\partial S_0} \le c|p|\|\tilde{f}\|_{1/2,p;\partial S_0;1/2,p;\partial S}, \tag{6.13}$$

$$\|\tilde{f}\|_{1/2,p;\partial S_0;1/2,p;\partial S} \le c|p|\|\mathcal{T}_p\tilde{f}\|_{-1/2,p;\partial S;-1/2,p;\partial S_0}. \tag{6.14}$$

Proof. Let $\tilde{\varphi}_+ = l_0\varphi_+$, and let $\tilde{\varphi}_- \in H_{1/2,p}(\partial S)$ be an extension of φ_- from ∂S_0 to ∂S such that $\tilde{\varphi}_+ - \tilde{\varphi}_- = \delta\tilde{\varphi} \in \overset{\circ}{H}_{1/2,p}(\partial S_0)$. We take $v_\pm = l^\pm\tilde{\varphi}_\pm$. Clearly, $v = (v_+, v_-) \in H_{1,p}(S)$ and

$$\|v\|_{1,p;S} \le c\|\tilde{\varphi}\|_{1/2,p;\partial S_0;1/2,p;\partial S}. \tag{6.15}$$

By (6.12) and (6.15),

$$|\langle \mathcal{T}_p\tilde{f}, \tilde{\varphi}\rangle_{0;\partial S_0}| \le c\|u\|_{1,p;S}\|v\|_{1,p;S}$$

$$\le c\|u\|_{1,p;S}\|\tilde{\varphi}\|_{1/2,p;\partial S_0;1/2,p;\partial S};$$

therefore, $\mathcal{T}_p\tilde{f} \in \overset{\circ}{H}_{-1/2,p}(\partial S_0) \times H_{-1/2,p}(\partial S_0)$ and

$$\|\mathcal{T}_p\tilde{f}\|_{-1/2,p;\partial S;-1/2,p;\partial S_0} \le c\|u\|_{1,p;S}.$$

Formula (6.7) now implies that

$$\|\mathcal{T}_p\tilde{f}\|_{-1/2,p;\partial S;-1/2,p;\partial S_0} \le c|p|\|\tilde{f}\|_{1/2,p;\partial S_0;1/2,p;\partial S}.$$

Taking in $\tilde{\varphi} = \tilde{f}$ and $v = u$ in (6.12) and separating the real and imaginary parts, we find that for any $p \in \bar{\mathbb{C}}_\kappa$,

$$\|u\|_{1,p;S}^2 \le c|p|\,|\langle T_p\tilde{f}, \tilde{f}\rangle_{0;\partial S_0}|; \tag{6.16}$$

hence,

$$\|u\|_{1,p;S} \le c|p|\,\|T_p\tilde{f}\|_{-1/2,p;\partial S;-1/2,p;\partial S_0}. \tag{6.17}$$

By the trace theorem and (6.17),

$$\|\tilde{f}\|_{1/2,p;\partial S_0;1/2,p;\partial S} \le c\|u\|_{1,p;S} \le c|p|\,\|T_p\tilde{f}\|_{-1/2,p;\partial S;-1/2,p;\partial S_0},$$

which proves (6.13) and (6.14).

If the range of T_p is not dense in $\mathring{H}_{-1/2,p}(\partial S_0) \times H_{-1/2,p}(\partial S_0)$, then there is a nonzero $\tilde{\varphi} \in H_{1/2,p}(\partial S_0) \times \mathring{H}_{1/2,p}(\partial S_0)$ such that

$$\langle T_p\tilde{f}, \tilde{\varphi}\rangle_{0;\partial S_0} = 0 \quad \forall \tilde{f} \in H_{1/2,p}(\partial S_0) \times \mathring{H}_{1/2,p}(\partial S_0). \tag{6.18}$$

Taking $\tilde{f} = \tilde{\varphi}$ in (6.18), we use (6.16) and the trace theorem to conclude that $\tilde{\varphi} = 0$. This contradiction completes the proof. □

We now introduce the operators $\hat{T}$, $\hat{T}^{-1}$ and T, T^{-1} in the usual way (see §3.1).

6.4 Theorem. *For any $\kappa > 0$ and $k \in \mathbb{R}$, the operator T is continuous and injective from the space $H_{1/2,k,\kappa}(\Gamma_0) \times \mathring{H}_{1/2,k,\kappa}(\Gamma_0)$ to $\mathring{H}_{-1/2,k-1,\kappa}(\Gamma_0) \times H_{-1/2,k-1,\kappa}(\Gamma_0)$, and its range is dense in $\mathring{H}_{-1/2,k-1,\kappa}(\Gamma_0) \times H_{-1/2,k-1,\kappa}(\Gamma_0)$. The inverse operator T^{-1}, extended by continuity from the range of T to $\mathring{H}_{-1/2,k,\kappa}(\Gamma_0) \times H_{-1/2,k,\kappa}(\Gamma_0)$, is continuous and injective from the space $\mathring{H}_{-1/2,k,\kappa}(\Gamma_0) \times H_{-1/2,k,\kappa}(\Gamma_0)$ to $H_{1/2,k-1,\kappa}(\Gamma_0) \times \mathring{H}_{1/2,k-1,\kappa}(\Gamma_0)$ for any $k \in \mathbb{R}$, and its range is dense in $H_{1/2,k-1,\kappa}(\Gamma_0) \times \mathring{H}_{1/2,k-1,\kappa}(\Gamma_0)$.*

This assertion is proved in the same way as those of a similar nature in the preceding chapters.

6.3 Time-dependent Potentials

Let $\hat{\alpha} \in \mathring{H}_{-1/2,p}(\partial S_0)$, $p \in \mathbb{C}_0$, and let $V_p\hat{\alpha}$ be the single-layer potential of density $\hat{\alpha}$. It is obvious that $V_p\hat{\alpha} \in H_{1,p}(S)$ and that the boundary operator $V_{p,0}$ generated by $V_p\hat{\alpha}$ on ∂S_0 through the equality

$$V_{p,0}\hat{\alpha} = \gamma_0^+ V_p\hat{\alpha} = \gamma_0^- V_p\hat{\alpha}$$

is continuous from $\overset{\circ}{H}_{-1/2,p}(\partial S_0)$ to $H_{1/2,p}(\partial S_0)$. Let

$$f_{\pm} = \gamma_0^{\mp} V_p \hat{\alpha}.$$

From the properties of single-layer potentials it follows that

$$f_+ = V_{p,0}\hat{\alpha}, \quad \delta f = f_+ - f_- = 0.$$

For this reason, in what follows we do not distinguish between the space $H_{1/2,p}(\partial S_0)$ and the subspace of $H_{1/2,p}(\partial S_0) \times \overset{\circ}{H}_{1/2,p}(\partial S_0)$ consisting of all the elements of the form $\{f_+, 0\}$, $f_+ \in H_{1/2,p}(\partial S_0)$.

6.5 Lemma. *For any $p \in C_0$, the operator $V_{p,0}$ is a homeomorphism from $\overset{\circ}{H}_{-1/2,p}(\partial S_0)$ to $H_{1/2,p}(\partial S_0)$, and for any function $\hat{\alpha} \in \overset{\circ}{H}_{-1/2,p}(\partial S_0)$, $p \in \bar{C}_\kappa$, $\kappa > 0$,*

$$\|V_{p,0}\hat{\alpha}\|_{1/2,p;\partial S_0} \le c|p|\|\hat{\alpha}\|_{-1/2,p;\partial S}, \tag{6.19}$$

$$\|\hat{\alpha}\|_{-1/2,p;\partial S} \le c|p|\|V_{p,0}\hat{\alpha}\|_{1/2,p;\partial S_0}. \tag{6.20}$$

Proof. Estimate (6.19) follows from Lemma 4.1. By the jump formula,

$$\mathcal{T}_p V_{p,0}\hat{\alpha} = \{\hat{\alpha}, \pi_0 \mathcal{T}_p^- \gamma^- \pi^- V_p \hat{\alpha}\} \quad \forall \hat{\alpha} \in \overset{\circ}{H}_{-1/2,p}(\partial S_0), \tag{6.21}$$

where, as before, $\mathcal{T}_p^{\pm}$ are the Poincaré–Steklov operators defined in relation to the domains $S^{\pm}$; hence,

$$\|\hat{\alpha}\|_{-1/2,p;\partial S} \le \|\mathcal{T}_p V_{p,0}\hat{\alpha}\|_{-1/2,p;\partial S;-1/2,p;\partial S_0}$$
$$\le c|p|\|V_{p,0}\hat{\alpha}\|_{1/2,p;\partial S_0}.$$

If the range of $V_{p,0}$ is not dense in $H_{1/2,p}(\partial S_0)$, then we can find a nonzero $\psi \in \overset{\circ}{H}_{-1/2,p}(\partial S_0)$ such that

$$(V_{p,0}\hat{\alpha}, \psi)_{0;\partial S_0} = 0 \quad \forall \hat{\alpha} \in \overset{\circ}{H}_{-1/2.p}(\partial S_0).$$

Taking $\hat{\alpha} = \psi$, from (6.16) with $u = V_p \psi$, (6.21), and the equality

$$V_{p,0}\hat{\alpha} = \{V_{p,0}\hat{\alpha}, 0\}$$

we see that

$$\|V_p \psi\|_{1,p;S}^2 \le c|p||\langle \mathcal{T}_p V_{p,0}\psi, V_{p,0}\psi \rangle_{0;\partial S_0}| = 0;$$

therefore, $V_p \psi = 0$, so $V_{p,0}\psi = 0$, which implies that $\psi = 0$. This contradiction completes the proof. $\qquad\square$

We now define the time-dependent (retarded) single-layer potential $V\alpha$ and its corresponding boundary operator V_0 in the usual way.

6.6 Theorem. *For any $\kappa > 0$ and $k \in \mathbb{R}$, the operator V_0 is continuous and injective from $\mathring{H}_{-1/2,k,\kappa}(\Gamma_0)$ to $H_{1/2,k-1,\kappa}(\Gamma_0)$, and its range is dense in $H_{1/2,k-1,\kappa}(\Gamma_0)$. The inverse V_0^{-1}, extended by continuity from the range of V_0, is continuous and injective from $H_{1/2,k,\kappa}(\Gamma_0)$ to $\mathring{H}_{-1/2,k-1,\kappa}(\Gamma_0)$ for any $k \in \mathbb{R}$, and its range is dense in $\mathring{H}_{-1/2,k-1,\kappa}(\Gamma_0)$. In addition, for any $\alpha \in \mathring{H}_{-1/2,k,\kappa}(\Gamma_0)$,*

$$\|V\alpha\|_{1,k-1,\kappa;G} \le c\|\alpha\|_{-1/2,k,\kappa;\Gamma}. \tag{6.22}$$

Proof. The assertions concerning V_0 and V_0^{-1} follow from Lemma 6.5, in particular, from (6.19) and (6.20).

Let $\hat{\alpha} \in \mathring{H}_{-1/2,p}(\partial S_0)$. By (6.16) with $u = V_p\hat{\alpha}$ and $\tilde{f} = \{V_{p,0}\hat{\alpha}, 0\}$,

$$\|V_p\hat{\alpha}\|_{1,p;S}^2 \le c|p| |(\hat{\alpha}, V_p\hat{\alpha})_{0;\partial S_0}| \le c|p|^2 \|\hat{\alpha}\|_{-1/2,p;\partial S}^2;$$

hence,

$$\|V_p\hat{\alpha}\|_{1,p;S} \le c|p| \|\hat{\alpha}\|_{-1/2,p;\partial S}, \tag{6.23}$$

and (6.22) follows from (6.23).   □

Next, we introduce the double-layer potential $W_p\hat{\beta}$ for $\hat{\beta} \in \mathring{H}_{1/2,p}(\partial S_0)$, define its corresponding boundary operators $W_{p,0}^{\pm}$ on ∂S_0 by

$$W_{p,0}^{\pm} = \pi_0 W_p^{\pm},$$

and introduce a boundary operator $W_{p,0}$ by setting

$$W_{p,0}\hat{\beta} = \{W_{p,0}^+\hat{\beta}, W_{p,0}^+\hat{\beta} - W_{p,0}^-\hat{\beta}\} = \{W_{p,0}^+\hat{\beta}, -\hat{\beta}\},$$

where we have used the jump property of $W_p\hat{\beta}$. By Lemma 4.5, for any $p \in \mathbb{C}_\kappa$, $\kappa > 0$,

$$\|W_{p,0}\hat{\beta}\|_{1/2,p;\partial S_0;1/2,p;\partial S} = \|W_{p,0}^+\hat{\beta}\|_{1/2,p;\partial S_0} + \|\hat{\beta}\|_{1/2,p;\partial S}$$

$$\le c|p|^2 \|\hat{\beta}\|_{1/2,p;\partial S}.$$

We now define an operator $N_{p,0}$ by writing

$$N_{p,0} = T_p W_{p,0}. \tag{6.24}$$

It is obvious that $N_{p,0}\hat{\beta} = \{0, \pi_0 N_p\hat{\beta}\}$ for any $\hat{\beta} \in \mathring{H}_{1/2,p}(\partial S_0)$, where

$$N_p = T_p^+ W_p^+ = T_p^- W_p^-.$$

In what follows we do not distinguish between the space $H_{-1/2,p}(\partial S_0)$ and the subspace of $\mathring{H}_{-1/2,p}(\partial S_0) \times H_{-1/2,p}(\partial S_0)$ consisting of all the elements of the form $\{0, g_-\}$, where $g_- \in H_{-1/2,p}(\partial S_0)$.

6.7 Lemma. *For any $p \in \mathbb{C}_0$, the operator $W_{p,0}$ is continuous from the space $\mathring{H}_{1/2,p}(\partial S_0)$ to $H_{1/2,p}(\partial S_0) \times \mathring{H}_{1/2,p}(\partial S_0)$, while the operator $N_{p,0}$ is a homeomorphism from $\mathring{H}_{1/2,p}(\partial S_0)$ to $H_{-1/2,p}(\partial S_0)$. In addition, for any $\hat{\beta} \in \mathring{H}_{1/2,p}(\partial S_0)$, $p \in \bar{\mathbb{C}}_\kappa$, $\kappa > 0$,*

$$\|W_{p,0}\hat{\beta}\|_{1/2,p;\partial S_0;1/2,p;\partial S} \leq c|p|^2\|\hat{\beta}\|_{1/2,p;\partial S}, \tag{6.25}$$

$$\|N_{p,0}\hat{\beta}\|_{-1/2,p;\partial S_0} \leq c|p|^3\|\hat{\beta}\|_{1/2,p;\partial S}, \tag{6.26}$$

$$\|\hat{\beta}\|_{1/2,p;\partial S} \leq c|p|\|N_{p,0}\hat{\beta}\|_{-1/2,p;\partial S_0}. \tag{6.27}$$

Proof. Estimate (6.25) has already been proved. Inequality (6.26) follows from (6.24), (6.25), and (6.13). To prove (6.27), we see that for any density $\hat{\beta} \in \mathring{H}_{1/2,p}(\partial S_0)$,

$$\|\hat{\beta}\|_{1/2,p;\partial S}^2 = \|\gamma_0^- W_p\hat{\beta} - \gamma_0^+ W_p\hat{\beta}\|_{1/2,p;\partial S}^2 \leq c\|W_p\hat{\beta}\|_{1,p;S}^2. \tag{6.28}$$

By (6.28) and (6.16) with $u = W_p\hat{\beta}$,

$$\|\hat{\beta}\|_{1/2,p;\partial S}^2 \leq c|p||(N_{p,0}\hat{\beta}, \hat{\beta})_{0;\partial S_0}| \leq c|p|\|N_{p,0}\hat{\beta}\|_{-1/2,p;\partial S_0}\|\hat{\beta}\|_{1/2,p;\partial S},$$

which proves (6.27).

If the range of $N_{p,0}$ is not dense in $H_{-1/2,p}(\partial S_0)$, then there is a nonzero $\psi \in \mathring{H}_{1/2,p}(\partial S_0)$ such that $(N_{p,0}\hat{\beta}, \psi)_{0;\partial S_0} = 0$ for any $\hat{\beta} \in \mathring{H}_{1/2,p}(\partial S_0)$. Taking $\hat{\beta} = \psi$, we use (6.16) with $u = W_p\psi$ and find that

$$\|W_p\psi\|_{1,p;S}^2 \leq c|p||(N_{p,0}\psi, \psi)_{0;\partial S_0}| = 0;$$

therefore, $W_p\psi = 0$, so

$$\psi = \gamma_0^- W_p\psi - \gamma_0^+ W_p\psi = 0.$$

This contradiction proves the lemma.   □

We now introduce the time-dependent (retarded) double-layer potential $W\beta$ and its corresponding boundary operators W_0, N_0, and N_0^{-1} in the usual way.

6.8 Theorem. *For any $\kappa > 0$ and $k \in \mathbb{R}$, the operator W_0 is continuous from $\mathring{H}_{1/2,k,\kappa}(\Gamma_0)$ to $H_{1/2,k-2,\kappa}(\Gamma_0) \times \mathring{H}_{1/2,k,\kappa}(\Gamma_0)$, while the operator N_0 is continuous and injective from $\mathring{H}_{1/2,k,\kappa}(\Gamma_0)$ to $H_{-1/2,k-3,\kappa}(\Gamma_0)$, and its range is dense in $H_{-1/2,k-3,\kappa}(\Gamma_0)$. The inverse N_0^{-1}, extended by continuity from the range of N_0, is continuous and injective from $H_{-1/2,k,\kappa}(\Gamma_0)$ to $\mathring{H}_{1/2,k-1,\kappa}(\Gamma_0)$ for any $k \in \mathbb{R}$, and its range is dense in $\mathring{H}_{1/2,k-1,\kappa}(\Gamma_0)$. In addition, for any $\beta \in \mathring{H}_{1/2,k,\kappa}(\Gamma_0)$,*

$$\|W\beta\|_{1,k-2,\kappa;G} \leq c\|\beta\|_{1/2,k,\kappa;\Gamma}.$$

Proof. The assertion follows from Lemma 6.7 and the Laplace transform estimate

$$\|W_p\hat{\beta}\|^2_{1,p;S} \leq c|p|\,|(N_{p,0}\hat{\beta}, \hat{\beta})_{0;\partial S_0}| \leq c|p|^4\|\hat{\beta}\|^2_{1/2,p;\partial S}. \qquad \square$$

6.4 Infinite Plate with a Crack

We seek the solution of problem (DKD) in the form

$$u(X) = (V\alpha)(X) + (W\beta)(X), \quad X \in G, \tag{6.29}$$

where α and β are unknown densities defined on ∂S_0. Representation (6.29) yields the system of boundary equations

$$V_0\alpha + W_0\beta = \tilde{f} = \{f_+, \delta f\}. \tag{6.30}$$

6.9 Theorem. *For any $\tilde{f} \in H_{1/2,k,\kappa}(\Gamma_0) \times \mathring{H}_{1/2,k,\kappa}(\Gamma_0)$, $\kappa > 0$, $k \in \mathbb{R}$, system (6.30) has a unique solution $\{\alpha, \beta\} \in \mathring{H}_{-1/2,k-1,\kappa}(\Gamma_0) \times \mathring{H}_{1/2,k,\kappa}(\Gamma_0)$, in which case u defined by (6.29) belongs to $H_{1,k-1,\kappa}(G)$. If $k \geq 1$, then u is the weak solution of problem (DKD).*

Proof. In terms of Laplace transforms, (6.30) takes the form

$$V_{p,0}\hat{\alpha} + W_{p,0}\hat{\beta} = \hat{f} = \{\hat{f}_+, \delta\hat{f}\}, \tag{6.31}$$

where $\hat{f}(x,p) = \mathcal{L}\tilde{f}(x,t)$. Since the second "component" of the left-hand side in (6.31) is $-\hat{\beta}$, we have $\hat{\beta} = -\delta\hat{f}$. By Lemma 6.3, (6.31) is equivalent to

$$\mathcal{T}_p V_{p,0}\hat{\alpha} + N_{p,0}\hat{\beta} = \mathcal{T}_p\hat{f} = \{\delta\hat{g}, \hat{g}_-\}. \tag{6.32}$$

The first "component" of the left-hand side in (6.32) is $\hat{\alpha}$; hence, $\hat{\alpha} = \delta\hat{g}$. Consequently, (6.31) has a unique solution $\{\hat{\alpha}, \hat{\beta}\} = \{\delta\hat{g}, -\delta\hat{f}\}$ and, by (6.13),

$$\|\hat{\alpha}\|_{-1/2,p;\partial S} \leq c|p|\|\hat{f}\|_{1/2,p;\partial S_0;1/2,p;\partial S}, \tag{6.33}$$

$$\|\hat{\beta}\|_{1/2,p;\partial S} \leq c\|\hat{f}\|_{1/2,p;\partial S_0;1/2,p;\partial S}. \tag{6.34}$$

If $\{\hat{\alpha}, \hat{\beta}\} \in \mathring{H}_{-1/2,p}(\partial S_0) \times \mathring{H}_{1/2,p}(\partial S_0)$ is the solution of (6.31), then, by (6.16) with $u = V_p\hat{\alpha} + W_p\hat{\beta}$ and (6.13),

$$\|V_p\hat{\alpha} + W_p\hat{\beta}\|_{1,p;S}^2 \leq c|p|\,|\langle T_p\hat{f}, \hat{f}\rangle_{0;\partial S_0}|$$
$$\leq c|p|^2\|\hat{f}\|_{1/2,p;\partial S_0;1/2,p;\partial S}^2. \tag{6.35}$$

Using (6.33)–(6.35), we complete the proof in the standard way.   □

We now represent the solution of problem (DKN) in the form (6.29) and arrive at the system of boundary equations

$$TV_0\alpha + N_0\beta = \tilde{g} = \{\delta g, g_-\}. \tag{6.36}$$

6.10 Theorem. *For any* $\tilde{g} \in \mathring{H}_{-1/2,k,\kappa}(\Gamma_0) \times H_{-1/2,k,\kappa}(\Gamma_0)$, $\kappa > 0$, $k \in \mathbb{R}$, *system (6.36) has a unique solution* $\{\alpha, \beta\} \in \mathring{H}_{-1/2,k,\kappa}(\Gamma_0) \times \mathring{H}_{1/2,k-1,\kappa}(\Gamma_0)$, *in which case* u *defined by (6.29) belongs to* $H_{1,k-1,\kappa}(G)$. *If* $k \geq 1$, *then* u *is the weak solution of problem* (DKN).

Proof. Going over to Laplace transforms, we reduce (6.36) to the system

$$T_pV_{p,0}\hat{\alpha} + N_{p,0}\hat{\beta} = \hat{g} = \{\delta\hat{g}, \hat{g}_-\}, \tag{6.37}$$

where

$$\hat{g}(x, p) = \mathcal{L}\tilde{g}(x, t).$$

Comparing the first "components" on both sides of (6.37), we see that $\hat{\alpha} = \delta\hat{g}$. System (6.37) is equivalent to

$$V_{p,0}\hat{\alpha} + W_{p,0}\hat{\beta} = T_p^{-1}\hat{g} = \{\hat{f}_+, \delta\hat{f}\}. \tag{6.38}$$

From (6.38) it follows that $\hat{\beta} = -\delta\hat{f}$. Using (6.14), we obtain

$$\|\hat{\alpha}\|_{-1/2,p;\partial S} \leq \|\hat{g}\|_{-1/2,p;\partial S;-1/2,p;\partial S_0}, \tag{6.39}$$

$$\|\hat{\beta}\|_{1/2,p;\partial S} \leq c|p|\|\hat{g}\|_{-1/2,p;\partial S;-1/2,p;\partial S_0}. \tag{6.40}$$

If $\{\hat{\alpha}, \hat{\beta}\} \in \mathring{H}_{-1/2,p}(\partial S_0) \times \mathring{H}_{1/2,p}(\partial S_0)$ is the solution of (6.37), then, by (6.16) with $u = V_p\hat{\alpha} + W_p\hat{\beta}$ and (6.14),

$$\|V_p\hat{\alpha} + W_p\hat{\beta}\|^2_{1,p;S} \leq c|p||\langle \hat{g}, \mathcal{T}_p^{-1}\hat{g}\rangle_{0;\partial S_0}|$$

$$\leq c|p|^2\|\hat{g}\|^2_{-1/2,p;\partial S;-1/2,p;\partial S_0}. \tag{6.41}$$

Estimates (6.39)–(6.41) enable us to complete the proof by following the usual procedure. □

6.5 Finite Plate with a Crack

Let $\tilde{S}$ be a finite domain whose boundary is a simple closed C^2-curve ∂S_2, and let ∂S_0 be an open arc in $\tilde{S}$. The arc ∂S_0, which models a crack, is assumed to be part of a simple closed C^2-curve ∂S that lies strictly inside $\tilde{S}$. We denote by S_i^+ the domain interior to ∂S and by S^- the (infinite) domain exterior to ∂S_2. We also write

$$S = \tilde{S} \setminus \overline{\partial S_0}, \quad S_e^+ = \tilde{S} \setminus \bar{S}_i^+,$$

$$\partial S_1 = \partial S \setminus \overline{\partial S_0}, \quad \partial\check{S} = \partial S_0 \cup \partial S_2,$$

$$\tilde{G} = \tilde{S} \times (0, \infty), \quad G = S \times (0, \infty),$$

$$G_i^+ = S_i^+ \times (0, \infty), \quad G_e^+ = S_e^+ \times (0, \infty), \quad G^- = S^- \times (0, \infty),$$

$$\check{\Gamma} = \partial\check{S} \times (0, \infty).$$

The operators of restriction from $\mathbb{R}^2$ (or $S_i^+ \cup S_e^+ \cup S^-$) to S_i^+, S_e^+, $\tilde{S}$, S, and S^- are denoted by π_i^+, π_e^+, $\tilde{\pi}$, π, and π^-, respectively.

We use the symbols $\gamma_{\nu,i}^+$, $\gamma_{\nu,e}^+$, $\nu = 0, 1$, for the operators of trace on ∂S_ν (Γ_ν) from inside S_i^+ (Γ_i^+) and S_e^+ (Γ_e^+), and $\gamma_{2,e}^+$ and γ_2^- for the operators of trace on ∂S_2 (Γ_2) from inside S_e^+ (Γ_e^+) and S^- (Γ^-).

$H_{1,p}(S)$, $p \in \mathbb{C}$, is the space of all functions $u(x, p) = \{u_{+,i}, u_{+,e}\}$ such that

$$u_{+,i} \in H_{1,p}(S_i^+), \quad u_{+,e} \in H_{1,p}(S_e^+), \quad \gamma_{1,i}^+ u_{+,i} = \gamma_{1,e}^+ u_{+,e},$$

equipped with the norm

$$\|u\|_{1,p;S} = \|u_{+,i}\|_{1,p;S_i^+} + \|u_{+,e}\|_{1,p;S_e^+}.$$

To simplify the notation, if u is a function defined on S (or $\tilde{S}$) and $u_{+,i}$ and $u_{+,e}$ are its restrictions to S_i^+ and S_e^+, then we write

$$\gamma_{\nu,i}^+ u_{+,i} = \gamma_{\nu,i}^+ u, \quad \gamma_{\nu,e}^+ u_{+,e} = \gamma_{\nu,e}^+ u, \quad \nu = 0, 1, \quad \gamma_{2,e}^+ u_{+,e} = \gamma_{2,e}^+ u.$$

$\mathring{H}_{1,p}(S)$ is the subspace of $H_{1,p}(\tilde{S})$ consisting of all u such that

$$\gamma_{0,i}^+ u = \gamma_{0,e}^+ u = \gamma_{2,e}^+ u = 0.$$

$H_{1/2,p}(\partial\check{S}) = H_{1/2,p}(\partial S_0) \times \mathring{H}_{1/2,p}(\partial S_0) \times H_{1/2,p}(\partial S_2)$ is the space of functions of the form

$$\check{f} = \{\tilde{f}, f_2\}, \quad \tilde{f} = \{f_{+,i}, \delta f\}, \quad \delta f = f_{+,i} - f_{+,e}.$$

$H_{-1/2,p}(\partial\check{S}) = \mathring{H}_{-1/2,p}(\partial S_0) \times H_{-1/2,p}(\partial S_0) \times H_{-1/2,p}(\partial S_2)$, with elements

$$\check{g} = \{\tilde{g}, g_2\}, \quad \tilde{g} = \{\delta g, g_{+,e}\}, \quad \delta g = g_{+,i} - g_{+,e}$$

is the dual of $H_{1/2,p}(\partial\check{S})$ with respect to the duality generated by the $L^2(\partial\check{S})$-inner product

$$\langle \check{f}, \check{g} \rangle_{0;\partial\check{S}} = \langle \tilde{f}, \tilde{g} \rangle_{0;\partial S_0} + (f_2, g_2)_{0;\partial S_2}.$$

The spaces $H_{\pm 1/2,k,\kappa}(\Gamma_0)$ and $\mathring{H}_{\pm 1/2,k,\kappa}(\Gamma_0)$, equipped with the norms $\|f_+\|_{\pm 1/2,k,\kappa;\Gamma_0}$ and $\|\delta f\|_{\pm 1/2,k,\kappa;\Gamma}$, as well as $H_{1,k,\kappa}(G)$ and $H_{\pm 1/2,k,\kappa}(\check{\Gamma})$ and their norms $\|u\|_{1,k,\kappa;G}$ and $\|\check{f}\|_{\pm 1/2,k,\kappa;\check{\Gamma}}$, are now defined in the usual way, with reference to the above spaces and with the same convention regarding the notation for the various traces of functions defined on G.

In this section we consider only one initial-boundary value problem and only one integral representation for its solution. The other problems and solution representations are treated analogously.

Thus, the variational version of problem (DKD) consists in finding a function $u \in H_{1,0,\kappa}(G)$ such that

$$\int_0^\infty \left[a_S(u, v) - (B^{1/2}\partial_t u, B^{1/2}\partial_t v)_{0;S} \right] dt = 0$$

$$\forall v \in C_0^\infty(S \times [0, \infty)), \qquad (6.42)$$

$$\gamma_{0,i}^+ u = f_{+,i}, \quad \gamma_{0,e}^- u = f_{+,e}, \quad \gamma_{2,e}^+ u = f_2,$$

where $f_{+,i}$ and $f_{+,e}$ are prescribed on Γ_0 and f_2 is prescribed on Γ_2.

6.11 Theorem. *For any given* $\check{f} = \{\tilde{f}, f_2\} \in H_{1/2,1,\kappa}(\check{\Gamma})$, $\kappa > 0$, *where* $\tilde{f} = \{f_{+,i}, \delta f\}$ *and* $\delta f = f_{+,i} - f_{+,e}$, *problem* (6.42) *has a unique solution* $u \in H_{1,0,\kappa}(G)$. *If* $\check{f} \in H_{1/2,k,\kappa}(\check{\Gamma})$, $k \in \mathbb{R}$, *then* $u \in H_{1,k-1,\kappa}(G)$ *and*

$$\|u\|_{1,k-1,\kappa;G} \leq c\big(\|f_{+,i}\|_{1/2,k,\kappa;\Gamma_0} + \|\delta f\|_{1/2,k,\kappa;\Gamma} + \|f_2\|_{1/2,k,\kappa;\Gamma_2}\big).$$

The proof is a repeat of that of Theorem 6.1, with the obvious changes.

Let $\check{f} = \{f_{+,i}, \delta f, f_2\}$, $\check{\varphi} = \{\varphi_{+,i}, \delta\varphi, \varphi_2\} \in H_{1/2,p}(\partial\check{S})$, $p \in \mathbb{C}_0$, and let $u \in H_{1,p}(S)$ be the solution of the variational problem

$$a_S(u,v) + p^2(B^{1/2}u, B^{1/2}v)_{0;S} = 0 \quad \forall v \in \mathring{H}_{1,p}(S),$$
$$\gamma_{0,i}^+ u = f_{+,i}, \quad \gamma_{0,e}^+ u = f_{+,e}, \quad \gamma_{2,e}^+ u = f_2,$$

where v is any element of $H_{1,p}(S)$ such that

$$\gamma_{0,i}^+ v = \varphi_{+,i}, \quad \gamma_{0,e}^+ v = \varphi_{+,e}, \quad \gamma_{2,e}^+ v = \varphi_2.$$

For any $p \in \mathbb{C}_0$, we define the Poincaré–Steklov operator $\mathcal{T}_p^+$ by means of the equality

$$\langle \mathcal{T}_p^+ \check{f}, \check{\varphi}\rangle_{0;\partial\check{S}} = p^2(B^{1/2}u, B^{1/2}v)_{0;S} + a_S(u,v).$$

6.12 Lemma. *For any* $p \in \mathbb{C}_0$, *the operator* $\mathcal{T}_p^+$ *is a homeomorphism from* $H_{1/2,p}(\partial\check{S})$ *to* $H_{-1/2,p}(\partial\check{S})$, *and for any* $\check{f} \in H_{1/2,p}(\partial\check{S})$, $p \in \bar{\mathbb{C}}_\kappa$, $\kappa > 0$,

$$\|\mathcal{T}_p^+ \check{f}\|_{-1/2,p;\partial\check{S}} \leq c|p|\|\check{f}\|_{1/2,p;\partial\check{S}},$$
$$\|\check{f}\|_{1/2,p;\partial\check{S}} \leq c|p|\|\mathcal{T}_p^+ \check{f}\|_{-1/2,p;\partial\check{S}},$$

where $\|\cdot\|_{1/2,p;\partial\check{S}}$ *and* $\|\cdot\|_{-1/2,p;\partial\check{S}}$ *are, respectively, the norms on* $H_{1/2,p}(\partial\check{S})$ *and* $H_{-1/2,p}(\partial\check{S})$.

This assertion is proved just like Lemma 6.3.

In what follows, we use the symbol $\mathcal{T}_p^-$ for the Poincaré–Steklov operator constructed in relation to the domain S^-.

Let $\hat{\alpha}_0 \in \mathring{H}_{-1/2,p}(\partial S_0)$ and $\hat{\alpha}_2 \in H_{-1/2,p}(\partial S_2)$. For any $p \in \mathbb{C}_0$, we define the single-layer potential $V_p\hat{\alpha}$ of density $\hat{\alpha} = \{\hat{\alpha}_0, \hat{\alpha}_2\} \in \mathring{H}_{-1/2,p}(\partial S_0) \times H_{-1/2,p}(\partial S_2)$ by

$$(V_p\hat{\alpha})(x,p) = (V_p^{(0)}\hat{\alpha}_0)(x,p) + (V_p^{(2)}\hat{\alpha}_2)(x,p), \quad x \in \mathbb{R}^2,$$

where $V_p^{(j)}\hat{\alpha}_j$, $j = 0,2$, are the single-layer potentials of densities $\hat{\alpha}_j$ constructed in terms of the parts ∂S_j of ∂S, respectively.

Since $\gamma_{0,i}^{+}V_{p}\hat{\alpha} = \gamma_{0,e}^{+}V_{p}\hat{\alpha}$, the corresponding boundary operator $V_{p,0}$ is defined by

$$V_{p,0}\hat{\alpha} = \{(V_{p}\hat{\alpha})_0, 0, (V_{p}\hat{\alpha})_2\},$$

where

$$(V_{p}\hat{\alpha})_0 = \gamma_{0,i}^{+}V_{p}\hat{\alpha} = \gamma_{0,e}^{+}V_{p}\hat{\alpha}, \quad (V_{p}\hat{\alpha})_2 = \gamma_{2,e}^{+}V_{p}\hat{\alpha}.$$

$V_{p,0}$ is continuous as a mapping from the space $\mathring{H}_{-1/2,p}(\partial S_0) \times H_{-1/2,p}(\partial S_2)$ to $H_{1/2,p}(\partial S_0) \times H_{1/2,p}(\partial S_2)$. Writing

$$T_{p}^{+}V_{p,0}\hat{\alpha} = \{\delta(T_{p}^{+}V_{p,0}\hat{\alpha}), (T_{p}^{+}V_{p,0}\hat{\alpha})_{+,e}, (T_{p}^{+}V_{p,0}\hat{\alpha})_2\},$$

we immediately see that the jump formula leads to

$$\hat{\alpha}_0 = \delta(T_{p}^{+}V_{p,0}\hat{\alpha}), \quad \hat{\alpha}_2 = (T_{p}^{+}V_{p,0}\hat{\alpha})_2 - T_{p}^{-}(V_{p,0}\hat{\alpha})_2, \tag{6.43}$$

where, according to the above notation,

$$(V_{p,0}\hat{\alpha})_2 = (V_{p}\hat{\alpha})_2 = \gamma_{2,e}^{+}V_{p}\hat{\alpha}.$$

In what follows, we identify $H_{1/2,p}(\partial S_0) \times H_{1/2,p}(\partial S_2)$ with the subspace of $H_{1/2,p}(\partial \breve{S})$ of all functions of the form $\breve{f} = \{f_{+,i}, 0, f_2\}$ with $f_{+,i} \in H_{1/2,p}(\partial S_0)$ and $f_2 \in H_{1/2,p}(\partial S_2)$. We define the norm of $\hat{\alpha} \in \mathring{H}_{-1/2,p}(\partial S_0) \times H_{-1/2,p}(\partial S_2)$ by

$$\|\hat{\alpha}\|_{-1/2,p;\partial S;-1/2,p;\partial S_2} = \|\hat{\alpha}_0\|_{-1/2,p;\partial S} + \|\hat{\alpha}_2\|_{-1/2,p;\partial S_2}.$$

6.13 Lemma. *For any $p \in \mathbb{C}_0$, the operator $V_{p,0}$ is a homeomorphism from the space $\mathring{H}_{-1/2,p}(\partial S_0) \times H_{-1/2,p}(\partial S_2)$ to $H_{1/2,p}(\partial S_0) \times H_{1/2,p}(\partial S_2)$, and for any $\hat{\alpha} \in \mathring{H}_{-1/2,p}(\partial S_0) \times H_{-1/2,p}(\partial S_2)$, $p \in \bar{\mathbb{C}}_\kappa$, $\kappa > 0$,*

$$\|V_{p,0}\hat{\alpha}\|_{1/2,p;\partial \breve{S}} \le c|p|\|\hat{\alpha}\|_{-1/2,p;\partial S;-1/2,p;\partial S_2}, \tag{6.44}$$

$$\|\hat{\alpha}\|_{-1/2,p;\partial S;-1/2,p;\partial S_2} \le c|p|\|V_{p,0}\hat{\alpha}\|_{1/2,p;\partial \breve{S}}. \tag{6.45}$$

Proof. Let $\hat{\alpha} \in \mathring{H}_{-1/2,p}(\partial S_0) \times H_{-1/2,p}(\partial S_2)$. By the trace theorem,

$$\|V_{p,0}\hat{\alpha}\|_{1/2,p;\partial \breve{S}}^2 \le c(\|\pi V_{p}\hat{\alpha}\|_{1,p;S}^2 + \|\pi^{-}V_{p}\hat{\alpha}\|_{1,p;S-}^2). \tag{6.46}$$

Since in our case

$$\|\pi V_{p}\hat{\alpha}\|_{1,p;S}^2 \le c\,\mathrm{Re}\,\{\bar{p}\langle T_{p}^{+}V_{p,0}\hat{\alpha}, V_{p,0}\hat{\alpha}\rangle_{0;\partial \breve{S}}\},$$

$$\|\pi^{-}V_{p}\hat{\alpha}\|_{1,p;S-}^2 \le c\,\mathrm{Re}\,\{\bar{p}\langle T_{p}^{-}(V_{p,0}\hat{\alpha})_2, (V_{p,0}\hat{\alpha})_2\rangle_{0;\partial S_2}\},$$

from (6.46) and (6.43) it follows that

$$\|V_{p,0}\hat{\alpha}\|^2_{1/2,p;\partial\check{S}} \le c|p|\{|(\hat{\alpha}_0, (V_p\hat{\alpha})_0)_{0;\partial S_0}| + |(\hat{\alpha}_2, (V_p\hat{\alpha})_2)_{0;\partial S_2}|\}$$

$$\le c|p|\|\hat{\alpha}\|_{-1/2,p;\partial S;-1/2,p;\partial S_2}\|V_{p,0}\hat{\alpha}\|_{1/2,p;\partial\check{S}}, \tag{6.47}$$

which implies that (6.44) holds.

Estimate (6.45) follows from equalities (6.43) and the properties of the Poincaré–Steklov operators.

If the range of $V_{p,0}$ is not dense in $H_{1/2,p}(\partial S_0) \times H_{1/2,p}(\partial S_2)$, then there is a nonzero $\psi = \{\psi_0, \psi_2\} \in \mathring{H}_{-1/2,p}(\partial S_0) \times H_{-1/2,p}(\partial S_2)$ such that

$$\left((V_p\hat{\alpha})_0, \psi_0\right)_{0;\partial S_0} + \left((V_p\hat{\alpha})_2, \psi_2\right)_{0;\partial S_2} = 0$$

$$\forall \hat{\alpha} \in \mathring{H}_{-1/2,p}(\partial S_0) \times H_{-1/2,p}(\partial S_2). \tag{6.48}$$

Taking $\hat{\alpha} = \psi$ in (6.48), we see that, by (6.47) and (6.48), $V_p\psi = 0$. This implies that $V_{p,0}\psi = 0$; hence, $\psi = 0$, and the assertion is proved. $\qquad\square$

In the usual way, we now construct the time-dependent single-layer potential $V\alpha$ of density $\alpha \in \mathring{H}_{-1/2,k,\kappa}(\Gamma_0) \times H_{-1/2,k,\kappa}(\Gamma_2)$ and its corresponding boundary operator V_0, and the double-layer potential $W_0\beta$ of density $\beta \in \mathring{H}_{1/2,k,\kappa}(\Gamma_0)$.

Let $(W_{p,0}\hat{\beta})(x, p)$ be the Laplace transform of $(W_0\beta)(X)$. We define the boundary operator $\check{W}_{p,0}$ acting on densities $\hat{\beta} \in \mathring{H}_{1/2,p}(\partial S_0)$ by setting

$$\check{W}_{p,0}\hat{\beta} = \{\gamma^+_{0,i}W_{p,0}\hat{\beta}, -\hat{\beta}, \gamma^+_{2,e}W_{p,0}\hat{\beta}\}.$$

By (6.25) and Theorem 6.8, $\check{W}_{p,0}$ is continuous from $\mathring{H}_{1/2,p}(\partial S_0)$ to $H_{1/2,p}(\partial\check{S})$ and for any $\hat{\beta} \in \mathring{H}_{1/2,p}(\partial S_0)$, $p \in \bar{\mathbb{C}}_\kappa$, $\kappa > 0$,

$$\|\check{W}_{p,0}\hat{\beta}\|_{1/2,p;\partial\check{S}} \le c|p|^2\|\hat{\beta}\|_{-1/2,p;\partial S}.$$

The corresponding boundary operator in the spaces of originals is denoted by the symbol $\check{W}_0$.

We seek the solution of (6.42) in the form

$$u(X) = (V\alpha)(X) + (W_0\beta)(X), \quad X \in G, \tag{6.49}$$

where α and β are unknown densities. Representation (6.49) leads to the system of boundary equations

$$V_0\alpha + \check{W}_0\beta = \check{f}, \tag{6.50}$$

where

$$\check{f} = \{\tilde{f}, f_2\}, \quad \tilde{f} = \{f_{+,i}, \delta f\}, \quad \delta f = f_{+,i} - f_{+,e}.$$

6.14 Theorem. *For any given $\check{f} \in H_{1/2,k,\kappa}(\check{\Gamma})$, $k \in \mathbb{R}$, $\kappa > 0$, system
(6.50) has a unique solution $\{\alpha, \beta\} \in \mathring{H}_{-1/2,k-1,\kappa}(\Gamma_0) \times H_{-1/2,k-1,\kappa}(\Gamma_2) \times \mathring{H}_{1/2,k,\kappa}(\Gamma_0)$, in which case u defined by (6.49) belongs to $H_{1,k-1,\kappa}(G)$. If
$k \geq 1$, then u is the weak solution of problem* (DKD).

Proof. In terms of Laplace transforms, (6.50) takes the form

$$V_{p,0}\hat{\alpha} + W_{p,0}\hat{\beta} = \hat{f}, \tag{6.51}$$

where $\hat{f} = \{\hat{f}_{+,i}, \delta\hat{f}, \hat{f}_2\}$ is the transform of f. Comparing the second "components" on both the sides in (6.51), we see that

$$\hat{\beta} = -\delta\hat{f}. \tag{6.52}$$

By Lemma 6.12, (6.51) is equivalent to the equation

$$\mathcal{T}_p^+ V_{p,0}\hat{\alpha} + \mathcal{T}_p^+ W_{p,0}\hat{\beta} = \hat{g}, \tag{6.53}$$

where

$$\hat{g} = \{\delta\hat{g}, \hat{g}_{+,e}, \hat{g}_2\} = \mathcal{T}_p^+ \hat{f}.$$

Since the first "component" on the left-hand side in (6.53) is equal to $\hat{\alpha}_0$, it follows that

$$\hat{\alpha}_0 = \delta\hat{g}. \tag{6.54}$$

Remarking that in our case the second equation (6.43) is written as

$$\hat{\alpha}_2 = (\mathcal{T}_p^+ V_{p,0}\hat{\alpha} + \mathcal{T}_p^+ W_{p,0}\hat{\beta})_2 - \mathcal{T}_p^- (V_{p,0}\hat{\alpha} + W_{p,0}\hat{\beta})_2,$$

we find that

$$\hat{\alpha}_2 = (\mathcal{T}_p^+ \hat{f})_2 - \mathcal{T}_p^- \hat{f}_2. \tag{6.55}$$

By (6.52), (6.54), (6.55), and Lemma 6.12,

$$\|\hat{\beta}\|_{1/2,p;\partial S} = \|\delta\hat{f}\|_{1/2,p;\partial S},$$

$$\|\hat{\alpha}_0\|_{-1/2,p;\partial S} \leq c|p|\|\hat{f}\|_{1/2,p;\partial\check{S}}, \tag{6.56}$$

$$\|\hat{\alpha}_2\|_{-1/2,p;\partial S_2} \leq c|p|\|\hat{f}\|_{1/2,p;\partial\check{S}}.$$

At the same time, we have

$$\|V_{p,0}\hat{\alpha} + W_{p,0}\hat{\beta}\|_{1,p;S}^2 \leq c|p|\|\langle\mathcal{T}_p^+\hat{f}, \hat{f}\rangle_{0;\partial\check{S}}| \leq c|p|^2\|\hat{f}\|_{1/2,p;\partial\check{S}}^2. \tag{6.57}$$

Taking (6.56) and (6.57) into account, we now complete the proof by following the standard procedure.   □

Initial-Boundary Value Problems with Other Types of Boundary Conditions

7.1 Mixed Boundary Conditions

As mentioned in §1.1, here we assume that the closed boundary curve ∂S consists of two parts ∂S_1 and ∂S_2 such that

$$\text{mes}\, \partial S_\nu > 0, \quad \nu = 1,2, \quad \partial S_1 \cap \partial S_2 = \emptyset, \quad \partial S = \overline{\partial S_1} \cup \overline{\partial S_2},$$

and write

$$\Gamma_\nu = \partial S_\nu \times (0, \infty).$$

We denote by π_ν, $\nu = 1,2$, the operators of restriction from ∂S to ∂S_ν (or from Γ to Γ_ν).

The operators of trace on ∂S_ν (Γ_ν) for functions defined on $S^\pm$ ($G^\pm$) are denoted by $\gamma_\nu^\pm$. We remark, in particular, that $\gamma_\nu^\pm$ map $H_{1,k,\kappa}(G^\pm)$ continuously onto $H_{1/2,k,\kappa}(\Gamma_\nu)$ for any $k \in \mathbb{R}$.

The notation regarding the single-layer and double-layer potentials and the boundary operators generated by them is the same as in Chapter 4.

In this section we consider the initial-boundary value problems $(\text{DM}^\pm)$ with mixed boundary data; in other words, we assume that the displacement field is prescribed on Γ_1 and that the moments and shear force are prescribed on Γ_2.

The classical problems $(\text{DM}^\pm)$ consist in finding $u \in C^2(G^\pm) \cap C^1(\bar{G}^\pm)$ that satisfy, respectively,

$$B(\partial_t^2 u)(X) + (Au)(X) = 0, \quad X \in G^\pm,$$
$$u(x, 0+) = (\partial_t u)(x, 0+) = 0, \quad x \in S^\pm,$$
$$u^\pm(X) = f_1(X), \quad X \in \Gamma_1,$$
$$(Tu)^\pm(X) = g_2(X), \quad X \in \Gamma_2.$$

We call $u \in H_{1,0,\kappa}(G^{\pm})$ weak solutions of $(\mathrm{DM}^{\pm})$ if they satisfy

$$\int\limits_0^{\infty} \left[a_{\pm}(u, v) - (B^{1/2}\partial_t u, B^{1/2}\partial_t v)_{0;S^{\pm}} \right] dt$$

$$= \pm \int\limits_0^{\infty} (g_2, \gamma_2^{\pm} v^{\pm})_{0;\partial S_2} \, dt \quad \forall v \in C_0^{\infty}(\bar{G}^{\pm}), \ \gamma_1^{\pm} v^{\pm} = 0, \quad (7.1)$$

$$\gamma_1^{\pm} u = f_1.$$

7.1 Theorem. *For any* $f_1 \in H_{1/2,1,\kappa}(\Gamma_1)$ *and* $g_2 \in H_{-1/2,1,\kappa}(\Gamma_2)$, $\kappa > 0$, *problems (7.1) have unique solutions* $u \in H_{1,0,\kappa}(G^{\pm})$. *If* $f_1 \in H_{1/2,k,\kappa}(\Gamma_1)$ *and* $g_2 \in H_{-1/2,k,\kappa}(\Gamma_2)$, $k \in \mathbb{R}$, *then* $u \in H_{1,k-1,\kappa}(G^{\pm})$ *and*

$$\|u\|_{1,k-1,\kappa;G^{\pm}} \leq c \big(\|f_1\|_{1/2,k,\kappa;\Gamma_1} + \|g_2\|_{-1/2,k,\kappa;\Gamma_2} \big).$$

Proof. We prove the statement for problem (DM^+); the case of (DM^-) is treated similarly. We begin by rewriting (7.1) in terms of Laplace transforms. Let $\mathring{H}_{1,p}(S^+, \partial S_{\nu})$ be the subspace in $H_{1,p}(S^+)$ of all u such that

$$\pi_{3-\nu}\gamma^+ u = 0, \quad \nu = 1, 2;$$

that is, $\gamma^+ u \in \mathring{H}_{1/2,p}(\partial S_{\nu})$. In the transform domain, (7.1) turns into the problem (M_p^+) of seeking $\hat{u} \in H_{1,p}(S^+)$ such that for any $p \in \mathbb{C}_{\kappa}$,

$$p^2(B^{1/2}\hat{u}, B^{1/2}v)_{0;S^+} + a_+(\hat{u}, v)$$

$$= (\hat{g}_2, \gamma_2^+ v)_{0;\partial S_2} \quad \forall v \in \mathring{H}_{1,p}(S^+, \partial S_2), \quad (7.2)$$

$$\gamma_1^+ \hat{u} = \hat{f}_1,$$

where $\hat{g}_2$ and $\hat{f}_1$ are the Laplace transforms of g_2 and f_1, respectively. Since for any $v \in \mathring{H}_{1,p}(S^+, \partial S_2)$,

$$|(\hat{g}_2, \gamma_2^+ v)_{0;\partial S_2}| \leq \|\hat{g}_2\|_{-1/2,p;\partial S_2} \|\gamma^+ v\|_{1/2,p;\partial S}$$

$$\leq c \|\hat{g}_2\|_{-1/2,p;\partial S_2} \|v\|_{1,p;S^+},$$

the form $(\hat{g}_2, \gamma_2^+ v)_{0;\partial S_2}$ defines a bounded antilinear (conjugate linear) functional on $\mathring{H}_{1,p}(S^+, \partial S_2)$; hence, it can be written as

$$(\hat{g}_2, \gamma_2^+ v)_{0;\partial S_2} = (q_2, v)_{0;S^+} \quad \forall v \in \mathring{H}_{1,p}(S^+, \partial S_2), \quad (7.3)$$

where q_2 belongs to the dual $\big[\mathring{H}_{1,p}(S^+, \partial S_2) \big]'$ of $\mathring{H}_{1,p}(S^+, \partial S_2)$ and

$$\|q_2\|_{[\mathring{H}_{1,p}(S^+, \partial S_2)]'} \leq c \|\hat{g}_2\|_{-1/2,p;\partial S_2}. \quad (7.4)$$

First, let $\hat{u}_0 \in \mathring{H}_{1,p}(S^+, \partial S_2)$ be the solution of our problem with $\hat{f}_1 = 0$. Taking (7.3) into account, in this case we rewrite (7.2) in the form

$$p^2(B^{1/2}\hat{u}_0, B^{1/2}v)_{0;S^+} + a_+(\hat{u}_0, v)$$

$$= (q_2, v)_{0;S^+} \quad \forall v \in \mathring{H}_{1,p}(S^+, \partial S_2). \tag{7.5}$$

The unique solvability of (7.5) and the estimate

$$\|\hat{u}_0\|_{1,p;S^+} \le c|p|\|q_2\|_{[\mathring{H}_{1,p}(S^+, \partial S_2)]'} \tag{7.6}$$

are established by standard arguments.

Let l_ν, $\nu = 1, 2$, be extension operators that map $H_{1/2,p}(\partial S_\nu)$ continuously to $H_{1/2,p}(\partial S)$. In the general problem, we consider a function $w = l^+ l_1 f_1 \in H_{1,p}(S^+)$, remark that

$$\|w\|_{1,p;S^+} \le c\|\hat{f}_1\|_{1/2,p;\partial S_1}, \tag{7.7}$$

and seek a solution $\hat{u}$ of (7.2) of the form $\hat{u} = \hat{u}_0 + \hat{w}$. Clearly, $\hat{u}_0 \in \mathring{H}_{1,p}(S^+, \partial S_2)$ satisfies

$$p^2(B^{1/2}\hat{u}_0, B^{1/2}v)_{0;S^+} + a_+(\hat{u}_0, v)$$

$$= (q_2, v)_{0;S^+} - p^2(B^{1/2}w, B^{1/2}v)_{0;S^+} - a_+(w, v)$$

$$\forall v \in \mathring{H}_{1,p}(S^+, \partial S_2). \tag{7.8}$$

By (7.7),

$$|p^2(B^{1/2}w, B^{1/2}v)_{0;S^+} + a_+(w, v)| \le c\|w\|_{1,p;S^+}\|v\|_{1,p;S^+}$$

$$\le c\|f_1\|_{1/2,p;\partial S_1}\|v\|_{1,p;S^+};$$

hence,

$$p^2(B^{1/2}w, B^{1/2}v)_{0;S^+} + a_+(w, v) = (q_1, v)_{0;S^+} \quad \forall v \in \mathring{H}_{1,p}(S^+, \partial S_2),$$

where $q_1 \in [\mathring{H}_{1,p}(S^+, \partial S_2)]'$ and

$$\|q_1\|_{[\mathring{H}_{1,p}(S^+, \partial S_2)]'} \le c\|\hat{f}_1\|_{1/2,p;\partial S_1}. \tag{7.9}$$

Equation (7.8) now takes the form

$$p^2(B^{1/2}\hat{u}_0, B^{1/2}v)_{0;S^+} + a_+(\hat{u}_0, v)$$

$$= (q_2 - q_1, v)_{0;S^+} \quad \forall v \in \mathring{H}_{1,p}(S^+, \partial S_2). \tag{7.10}$$

We have already shown that (7.10) has a unique solution $\hat{u}_0 \in \mathring{H}_{1,p}(S^+, \partial S_2)$. By (7.6), (7.4), and (7.9),

$$\|\hat{u}_0\|_{1,p;S^+} \le c|p|\|q_1 - q_2\|_{[\mathring{H}_{1,p}(S^+,\partial S_2)]'}$$
$$\le c|p|(\|\hat{f}_1\|_{1/2,p;\partial S_1} + \|\hat{g}_2\|_{-1/2,p;\partial S_2});$$

therefore, $\hat{u} = \hat{u}_0 + w$ satisfies

$$\|\hat{u}\|_{1,p;S^+} \le c|p|(\|\hat{f}_1\|_{1/2,p;\partial S_1} + \|\hat{g}_2\|_{-1/2,p;\partial S_2}). \tag{7.11}$$

The proof is now completed by means of standard arguments. $\qquad\square$

Next, we consider suitable integral representations for the solutions of $(DM^\pm)$. With this aim, for $p \in \mathbb{C}_0$, we introduce three pairs of boundary integral operators and study their properties.

Let $f \in H_{1/2,p}(\partial S)$, and let $\nu, \tau = 1, 2$, $\nu \ne \tau$. We define operators $\pi^\pm_{p,\nu\tau}$ by

$$\pi^\pm_{p,\nu\tau} f = \{\pi_\nu f, \pi_\tau \mathcal{T}^\pm_p f\},$$

where $\mathcal{T}^\pm_p$ are the Poincaré–Steklov operators constructed in terms of the domains $S^\pm$.

Let $g \in H_{-1/2,p}(\partial S)$. For the same values of ν and τ as above, the operators $\theta^\pm_{p,\nu\tau}$ are defined by

$$\theta^\pm_{p,\nu\tau} g = \pi^\pm_{p,\nu\tau}(\mathcal{T}^\pm_p)^{-1}g.$$

Finally, for $\{f_\nu, g_\tau\} \in H_{1/2,p}(\partial S_\nu) \times H_{-1/2,p}(\partial S_\tau)$, we define operators $\rho^\pm_{p,\tau\nu}$ by

$$\rho^\pm_{p,\tau\nu}\{f_\nu, g_\tau\} = \pi^\pm_{p,\tau\nu}(\pi^\pm_{p,\nu\tau})^{-1}\{f_\nu, g_\tau\}.$$

The norm of $\{f_\nu, g_\tau\}$ in $H_{1/2,p}(\partial S_\nu) \times H_{-1/2,p}(\partial S_\tau)$ is

$$\|\{f_\nu, g_\tau\}\|_{1/2,p;\partial S_\nu;-1/2,p;\partial S_\tau} = \|f_\nu\|_{1/2,p;\partial S_\nu} + \|g_\tau\|_{-1/2,p;\partial S_\tau}.$$

7.2 Lemma. (i) *The operators $\pi^\pm_{p,\nu\tau}$ are homeomorphisms from $H_{1/2,p}(\partial S)$ to $H_{1/2,p}(\partial S_\nu) \times H_{-1/2,p}(\partial S_\tau)$, and for any $f \in H_{1/2,p}(\partial S)$, $p \in \bar{\mathbb{C}}_\kappa$, $\kappa > 0$,*

$$\|\pi^\pm_{p,\nu\tau} f\|_{1/2,p;\partial S_\nu;-1/2,p;\partial S_\tau} \le c|p|\|f\|_{1/2,p;\partial S}, \tag{7.12}$$

$$\|f\|_{1/2,p;\partial S} \le c|p|\|\pi^\pm_{p,\nu\tau} f\|_{1/2,p;\partial S_\nu;-1/2,p;\partial S_\tau}. \tag{7.13}$$

(ii) *The operators $\theta^\pm_{p,\nu\tau}$ are homeomorphisms from $H_{-1/2,p}(\partial S)$ to the space $H_{1/2,p}(\partial S_\nu) \times H_{-1/2,p}(\partial S_\tau)$, and for any $g \in H_{-1/2,p}(\partial S)$, $p \in \bar{\mathbb{C}}_\kappa$, $\kappa > 0$,*

$$\|\theta^\pm_{p,\nu\tau} g\|_{1/2,p;\partial S_\nu;-1/2,p;\partial S_\tau} \le c|p|\|g\|_{-1/2,p;\partial S}, \tag{7.14}$$

$$\|g\|_{-1/2,p;\partial S} \le c|p|\|\theta^\pm_{p,\nu\tau} g\|_{1/2,p;\partial S_\nu;-1/2,p;\partial S_\tau}. \tag{7.15}$$

(iii) *The operators $\rho^{\pm}_{p,\nu\tau}$ are homeomorphisms from the space $H_{1/2,p}(\partial S_{\nu}) \times H_{-1/2,p}(\partial S_{\tau})$ to $H_{1/2,p}(\partial S_{\tau}) \times H_{-1/2,p}(\partial S_{\nu})$, and for any element $\{f_{\nu}, g_{\tau}\} \in H_{1/2,p}(\partial S_{\nu}) \times H_{-1/2,p}(\partial S)$, $p \in \bar{\mathbb{C}}_{\kappa}$, $\kappa > 0$,*

$$\|\rho^{\pm}_{p,\nu\tau}\{f_{\nu}, g_{\tau}\}\|_{1/2,p;\partial S_{\tau};-1/2,p;\partial S_{\nu}}$$
$$\leq c|p|\|\{f_{\nu}, g_{\tau}\}\|_{1/2,p;\partial S_{\nu};-1/2,p;\partial S_{\tau}}, \tag{7.16}$$

$$\|\{f_{\nu}, g_{\tau}\}\|_{1/2,p;\partial S_{\nu};-1/2,p;\partial S_{\tau}}$$
$$\leq c|p|\|\rho^{\pm}_{p,\nu\tau}\{f_{\nu}, g_{\tau}\}\|_{1/2,p;\partial S_{\tau};-1/2,p;\partial S_{\nu}}. \tag{7.17}$$

Proof. (i) The continuity of $\pi^{\pm}_{p,\nu\tau}$ and (7.12) follow from Lemma 3.1. Let $u \in H_{1,p}(S^{\pm})$ be the unique solutions of the problems

$$p^2(B^{1/2}u, B^{1/2}v)_{0;S^{\pm}} + a_{\pm}(u, v)$$
$$= \pm(g_{\tau}, \gamma^{\pm}_{\tau}v)_{0;\partial S_{\tau}} \quad \forall v \in \mathring{H}_{1,p}(S^{\pm}, \partial S_{\tau}), \tag{7.18}$$
$$\gamma^{\pm}_{\nu}u = f_{\nu},$$

where $\{f_{\nu}, g_{\tau}\} \in H_{1/2,p}(\partial S_{\nu}) \times H_{-1/2,p}(\partial S_{\tau})$. By (7.11),

$$\|u\|_{1,p;S^{\pm}} \leq c|p|\|\{f_{\nu}, g_{\tau}\}\|_{1/2,p;\partial S_{\nu};-1/2,p;\partial S_{\tau}}. \tag{7.19}$$

If $f = \gamma^{\pm}u \in H_{1/2,p}(\partial S)$, then $\pi^{\pm}_{p,\nu\tau}f = \{f_{\nu}, g_{\tau}\}$; hence, $\pi^{\pm}_{p,\nu\tau}$ are surjective. The trace theorem and (7.19) imply that (7.13) holds.

(ii) By Lemma 3.1 and (i) above, the operators $\theta^{\pm}_{p,\nu\tau}$ are homeomorphisms from $H_{-1/2,p}(\partial S)$ to $H_{1/2,p}(\partial S_{\nu}) \times H_{-1/2,p}(\partial S_{\tau})$. Let $u \in H_{1,p}(S^{\pm})$ be the unique solutions of the problems

$$p^2(B^{1/2}u, B^{1/2}v)_{0;S^{\pm}} + a_{\pm}(u, v) = \pm(g, \gamma^{\pm}v)_{0;\partial S} \quad \forall v \in H_{1,p}(S^{\pm}),$$

where $g \in H_{-1/2,p}(\partial S)$. By (3.12),

$$\|u\|_{1,p;S^{\pm}} \leq c|p|\|g\|_{-1/2,p;\partial S}.$$

If $f = (\mathcal{T}^{\pm}_p)^{-1}g$, then $\|f\|_{1/2,p;\partial S} \leq c|p|\|g\|_{-1/2,p;\partial S}$. We have

$$\|\pi_{\nu}f\|_{1/2,p;\partial S_{\nu}} + \|\pi_{\tau}g\|_{-1/2,p;\partial S_{\tau}} \leq \|f\|_{1/2,p;\partial S} + \|g\|_{-1/2,p;\partial S}$$
$$\leq c|p|\|g\|_{-1/2,p;\partial S},$$

which proves (7.14).

Now let u be the solution of (7.2) with boundary data f_1 and g_2, and let $\gamma^{\pm}u = f$ and $\mathcal{T}^{\pm}_p f = g$. From (3.5) and (7.11) it follows that (7.15) holds.

(iii) The definition of the $\rho^{\pm}_{p,\nu\tau}$ and statement (i) above imply that these operators are homeomorphisms from the space $H_{1/2,p}(\partial S_{\nu}) \times H_{-1/2,p}(\partial S_{\tau})$

to $H_{1/2,p}(\partial S_\tau) \times H_{-1/2,p}(\partial S_\nu)$. As above, let u be the solution of (7.2) with boundary data f_1 and g_2, and let $f = \gamma^\pm u$ and $g = T_p^\pm f$. Then

$$\|\pi_\nu g\|_{-1/2,p;\partial S_\nu} + \|\pi_\tau f\|_{1/2,p;\partial S_\tau} \leq \|g\|_{-1/2,p;\partial S} + \|f\|_{1/2,p;\partial S} \leq c\|u\|_{1,p;S^\pm}$$

$$\leq c|p| \| \{f_\nu, g_\tau\} \|_{1/2,p;\partial S_\nu;-1/2,p;\partial S_\tau},$$

which proves (7.16). Estimate (7.17) is established similarly.   □

We now consider four representations for the solutions of problems $(\mathrm{DM}^\pm)$ in terms of time-dependent (retarded) plate potentials and demonstrate the unique solvability of the corresponding systems of boundary equations. First, we seek these solutions in the form

$$u(X) = (V\alpha)(X), \quad X \in G^\pm, \tag{7.20}$$

or $u = \pi^\pm V\alpha$, where $V\alpha$ is the single-layer potential of unknown density α. This leads to the system of boundary equations

$$\begin{aligned} (V\alpha)(X) &= f_1(X), \quad X \in \Gamma_1, \\ (T^\pm V\alpha)(X) &= g_2(X), \quad X \in \Gamma_2, \end{aligned} \tag{7.21}$$

or

$$\pi_1 V_0 \alpha = f_1, \quad \pi_2 T^\pm V_0 \alpha = g_2,$$

where V_0 is the boundary operator generated by $V\alpha$.

7.3 Theorem. *For any given $f_1 \in H_{1/2,k,\kappa}(\Gamma_1)$ and $g_2 \in H_{-1/2,k,\kappa}(\Gamma_2)$, $\kappa > 0$, $k \in \mathbb{R}$, systems (7.21) have unique solutions $\alpha \in H_{-1/2,k-2,\kappa}(\Gamma)$, in which case the functions u defined by (7.20) belong, respectively, to the spaces $H_{1,k-1,\kappa}(G^\pm)$. If $k \geq 1$, then these functions are the weak solutions of problems $(\mathrm{DM}^\pm)$.*

Proof. For simplicity, we prove this only for (DM^+); the case of (DM^-) is treated similarly.

In terms of Laplace transforms, (7.21) becomes

$$\pi_1 V_{p,0} \hat{\alpha} = \hat{f}_1, \quad \pi_2 T_p^+ V_{p,0} \hat{\alpha} = \hat{g}_2,$$

or

$$\pi_{p,12}^+ V_{p,0} \hat{\alpha} = \{\hat{f}_1, \hat{g}_2\}. \tag{7.22}$$

By Lemmas 7.2 and 4.1, system (7.22) has a unique solution $\hat{\alpha} \in H_{-1/2,p}(\partial S)$, which, for any $p \in \mathbb{C}_\kappa$, satisfies

$$\|\hat{\alpha}\|_{-1/2,p;\partial S} \leq c|p|^2 \| \{\hat{f}_1, \hat{g}_2\} \|_{1/2,p;\partial S_1;-1/2,p;\partial S_2}. \tag{7.23}$$

Taking (7.23) and Theorem 7.1 into account, we now complete the proof by following the standard procedure.   □

The second solution representation is

$$u(X) = (W\beta)(X), \quad X \in G^{\pm}, \tag{7.24}$$

or $u = \pi^{\pm}W\beta$, where $W\beta$ is the double-layer potential of unknown density β. This leads to the system of boundary integral equations

$$(W^{\pm}\beta)(X) = f_1(X), \quad X \in \Gamma_1, \tag{7.25}$$
$$(N\beta)(X) = g_2(X), \quad X \in \Gamma_2,$$

or

$$\pi_1 W^{\pm}\beta = f_1, \quad \pi_2 N\beta = g_2,$$

where $W^{\pm}$ are the boundary operators generated by $W\beta$ and

$$N = \mathcal{T}^+W^+ = \mathcal{T}^-W^-.$$

7.4 Theorem. *For any given $f_1 \in H_{1/2,k,\kappa}(\Gamma_1)$ and $g_2 \in H_{-1/2,k,\kappa}(\Gamma_2)$, $\kappa > 0$, $k \in \mathbb{R}$, systems (7.25) have unique solutions $\beta \in H_{1/2,k-2,\kappa}(\Gamma)$, in which case the functions u defined by (7.24) belong, respectively, to the spaces $H_{1,k-1,\kappa}(G^{\pm})$. If $k \geq 1$, then these functions are the weak solutions of problems* (DM$^{\pm}$).

Proof. Once again, for simplicity we discuss only the interior problem. In terms of Laplace transforms, system (7.25) for (DM$^+$) takes the form

$$\pi_1 W_p^+\hat{\beta} = \hat{f}_1, \quad \pi_2 N_p\hat{\beta} = \hat{g}_2,$$

or

$$\pi_{p,12}^+ W_p^+\hat{\beta} = \{\hat{f}_1, \hat{g}_2\}. \tag{7.26}$$

Let $\hat{g} = N_p\hat{\beta}$. Then $W_p^+\hat{\beta} = (\mathcal{T}_p^+)^{-1}\hat{g}$, and (7.26) can be written as

$$\theta_{p,12}^+\hat{g} = \{\hat{f}_1, \hat{g}_2\}. \tag{7.27}$$

By Lemma 7.2 and Theorem 4.8, equation (7.27)—hence, also (7.26)—has a unique solution $\hat{\beta} \in H_{1/2,p}(\partial S)$, which satisfies

$$\|\hat{\beta}\|_{1/2,p;\partial S} \leq c|p|^2 \|\{\hat{f}_1, \hat{g}_2\}\|_{1/2,p;\partial S_1;-1/2,p;\partial S_2}. \tag{7.28}$$

We now complete the proof in the usual way, making use of Theorem 7.1 and (7.28). □

The third integral representation is

$$u(X) = (V\alpha_1)(X) + (W\beta_2)(X), \quad X \in G^{\pm}, \tag{7.29}$$

or

$$u = \pi^{\pm}(V\alpha_1 + W\beta_2),$$

where $\alpha_1 \in \mathring{H}_{-1/2,k,\kappa}(\Gamma_1)$ and $\beta_2 \in \mathring{H}_{1/2,k,\kappa}(\Gamma_2)$ are unknown densities. This yields the systems of boundary equations

$$(V_0\alpha_1)(X) + (W^\pm\beta_2)(X) = f_1(X), \quad X \in \Gamma_1,$$
$$(\mathcal{T}^\pm V_0\alpha_1)(X) + (N\beta_2)(X) = g_2(X), \quad X \in \Gamma_2, \tag{7.30}$$

or

$$\pi_1(V_0\alpha_1 + W^\pm\beta_2) = f_1, \quad \pi_2(\mathcal{T}^\pm V_0\alpha_1 + N\beta_2) = g_2.$$

7.5 Theorem. *For any $f_1 \in H_{1/2,k,\kappa}(\Gamma_1)$ and $g_2 \in H_{-1/2,k,\kappa}(\Gamma_2)$, $\kappa > 0$, $k \in \mathbb{R}$, systems (7.30) have unique solutions $\{\alpha_1, \beta_2\} \in \mathring{H}_{-1/2,k-1,\kappa}(\Gamma_1) \times \mathring{H}_{1/2,k-1,\kappa}(\Gamma_2)$, in which case the functions u defined by (7.29) belong, respectively, to $H_{1,k-1,\kappa}(G^\pm)$. If $k \geq 1$, then these functions are the weak solutions of problems (DM$^\pm$).*

Proof. We discuss only the interior problem; the exterior one is dealt with similarly. In terms of Laplace transforms, (7.30) for (DM$^+$) becomes

$$\pi^+_{p,12}(V_{p,0}\hat{\alpha}_1 + W^+_p\hat{\beta}_2) = \{\hat{f}_1, \hat{g}_2\}. \tag{7.31}$$

We claim that

$$\hat{\alpha}_1 = [(\theta^+_{p,12})^{-1} - (\theta^-_{p,12})^{-1}]\{\hat{f}_1, \hat{g}_2\},$$
$$\hat{\beta}_2 = [(\pi^-_{p,12})^{-1} - (\pi^+_{p,12})^{-1}]\{\hat{f}_1, \hat{g}_2\}$$

is the solution of (7.31). True, from the definition of $\pi^\pm_{p,12}$ it follows that $\pi_1[(\pi^+_{p,12})^{-1} - (\pi^-_{p,12})^{-1}] = 0$, so $\hat{\beta}_2 \in \mathring{H}_{1/2,p}(\partial S_2)$. Similarly, it is shown that $\hat{\alpha}_1 \in \mathring{H}_{-1/2,p}(\partial S_1)$. Consequently, since

$$(\theta^\pm_{p,12})^{-1} = \mathcal{T}^\pm_p(\pi^\pm_{p,12})^{-1}, \quad V_{p,0}(\mathcal{T}^+_p - \mathcal{T}^-_p) = I,$$

where I is the identity operator, we have

$$V_{p,0}\hat{\alpha}_1 + W^+_p\hat{\beta}_2$$
$$= V_{p,0}\{(\theta^+_{p,12})^{-1} - (\theta^-_{p,12})^{-1} + \mathcal{T}^-_p[(\pi^-_{p,12})^{-1} - (\pi^+_{p,12})^{-1}]\}\{\hat{f}_1, \hat{g}_2\}$$
$$= V_{p,0}[(\theta^+_{p,12})^{-1} - \mathcal{T}^-_p(\pi^+_{p,12})^{-1}]\{\hat{f}_1, \hat{g}_2\} = (\pi^+_{p,12})^{-1}\{\hat{f}_1, \hat{g}_2\},$$

which proves the assertion.

If (7.31) has more than one solution, then the difference $\{\tilde{\alpha}_1, \tilde{\beta}_2\}$ of any two solutions belongs to $\in \mathring{H}_{-1/2,p}(\partial S_1) \times \mathring{H}_{1/2,p}(\partial S_2)$ and satisfies

$$\pi^+_{p,12}(V_{p,0}\tilde{\alpha}_1 + W^+_p\tilde{\beta}_2) = \{0, 0\},$$

so

$$V_{p,0}\tilde{\alpha}_1 + W^+_p\tilde{\beta}_2 = 0.$$

It is clear that $\pi^{\pm}(V_p\tilde{\alpha}_1 + W_p\tilde{\beta}_2)$ are solutions of $(\mathrm{M}_p^{\pm})$ with $\hat{f}_1 = 0$ and $\hat{g}_2 = 0$; consequently,

$$\pi^{\pm}(V_p\tilde{\alpha}_1 + W_p\tilde{\beta}_2) = 0.$$

This implies that

$$\tilde{\beta}_2 = \gamma^-(V_p\tilde{\alpha}_1 + W_p\tilde{\beta}_2) - \gamma^+(V_p\tilde{\alpha}_1 + W_p\tilde{\beta}_2) = 0,$$
$$\tilde{\alpha}_1 = \mathcal{T}_p^+\gamma^+(V_p\tilde{\alpha}_1 + W_p\tilde{\beta}_2) - \mathcal{T}_p^-\gamma^-(V_p\tilde{\alpha}_1 + W_p\tilde{\beta}_2) = 0,$$

which proves that the solution of (7.31) is unique.

The estimates

$$\|\hat{\alpha}_1\|_{-1/2,p;\partial S} \le c|p|\|\{\hat{f}_1, \hat{g}_2\}\|_{1/2,p;\partial S_1;-1/2,p;\partial S_2},$$
$$\|\hat{\beta}_2\|_{1/2,p;\partial S} \le c|p|\|\{\hat{f}_1, \hat{g}_2\}\|_{1/2,p;\partial S_1;-1/2,p;\partial S_2}$$

follow from (7.13) and (7.15). The proof now continues according to the standard procedure. □

The fourth representation is

$$u(X) = (W\beta_1)(X) + (V\alpha_2)(X), \quad X \in G^{\pm}, \tag{7.32}$$

or

$$u = \pi^{\pm}(W\beta_1 + V\alpha_2),$$

where $\beta_1 \in \mathring{H}_{1/2,k,\kappa}(\Gamma_1)$ and $\alpha_2 \in \mathring{H}_{-1/2,k,\kappa}(\Gamma_2)$. This representation leads to the boundary systems

$$(W^{\pm}\beta_1)(X) + (V_0\alpha_2)(X) = f_1(X), \quad X \in \Gamma_1,$$
$$(N\beta_1)(X) + (\mathcal{T}^{\pm}V_0\alpha_2)(X) = g_2(X), \quad X \in \Gamma_2, \tag{7.33}$$

or

$$\pi_1(W^{\pm}\beta_1 + V_0\alpha_2) = f_1, \quad \pi_2(N\beta_1 + \mathcal{T}^{\pm}V_0\alpha_2) = g_2.$$

7.6 Theorem. *For any $f_1 \in H_{1/2,k,\kappa}(\Gamma_1)$ and $g_2 \in H_{-1/2,k,\kappa}(\Gamma_2)$, $\kappa > 0$, $k \in \mathbb{R}$, systems (7.33) have unique solutions $\{\beta_1, \alpha_2\} \in \mathring{H}_{1/2,k-2,\kappa}(\Gamma_1) \times \mathring{H}_{-1/2,k-2,\kappa}(\Gamma_2)$, in which case the functions u defined by (7.32) belong, respectively, to $H_{1,k-1,\kappa}(G^{\pm})$. If $k \ge 1$, then these functions are the weak solutions of problems $(\mathrm{DM}^{\pm})$.*

Proof. As for the preceding assertions, we consider only the interior problem. In the Laplace transform domain, (7.33) for (DM^+) becomes

$$\pi_{p,12}^+(W_p^+\hat{\beta}_1 + V_{p,0}\hat{\alpha}_2) = \{\hat{f}_1, \hat{g}_2\}. \tag{7.34}$$

We claim that

$$\hat{\beta}_1 = \left[(\pi^-_{p,12})^{-1}\rho^+_{p,21} - (\pi^+_{p,12})^{-1}\right]\{\hat{f}_1, \hat{g}_2\},$$

$$\hat{\alpha}_2 = \left[(\theta^+_{p,12})^{-1} - (\theta^-_{p,12})^{-1}\rho^+_{p,21}\right]\{\hat{f}_1, \hat{g}_2\}$$

is the solution of (7.34). True, we have

$$W^+_p\hat{\beta}_1 + V_{p,0}\hat{\alpha}_2$$

$$= V_{p,0}\Big\{T^-_p\left[(\pi^-_{p,12})^{-1}\rho^+_{p,21} - (\pi^+_{p,12})^{-1}\right]$$

$$+ (\theta^+_{p,12})^{-1} - (\theta^-_{p,12})^{-1}\rho^+_{p,21}\Big\}\{\hat{f}_1, \hat{g}_2\}$$

$$= V_{p,0}\left[T^+_p(\pi^+_{p,12})^{-1} - T^-_p(\pi^+_{p,12})^{-1}\right]\{\hat{f}_1, \hat{g}_2\} = (\pi^+_{p,12})^{-1}\{\hat{f}_1, \hat{g}_2\},$$

so (7.34) holds.
   Since

$$\pi_2\left[(\pi^-_{p,12})^{-1}\rho^+_{p,21} - (\pi^+_{p,12})^{-1}\right]\{\hat{f}_1, \hat{g}_2\} = 0,$$

it follows that $\hat{\beta}_1 \in \mathring{H}_{p,1/2}(\partial S_1)$. Similarly, we verify that $\hat{\alpha} \in \mathring{H}_{-1/2,p}(\partial S)$. To establish the unique solvability of (7.34), we repeat the arguments in the proof of Theorem 7.5, with the obvious changes.
   By (7.13), (7.15), and (7.16), for any $p \in \mathbb{C}_\kappa$,

$$\|\hat{\beta}_1\|_{1/2,p;\partial S} \leq c|p|^2\|\{\hat{f}_1, \hat{g}_2\}\|_{1/2,p;\partial S_1;-1/2,p;\partial S_2},$$

$$\|\hat{\alpha}_2\|_{-1/2,p;\partial S} \leq c|p|^2\|\{\hat{f}_1, \hat{g}_2\}\|_{1/2,p;\partial S_1;-1/2,p;\partial S_2}.$$

The rest of the proof is now completed according to the standard scheme. □

7.2 Combined Boundary Conditions

In the classical dynamic problems $(DC^\pm_1)$ with combined boundary conditions, we seek functions $u = (\bar{u}, u_3) \in C^2(G^\pm) \cap C^1(\bar{G}^\pm)$ that satisfy

$$B(\partial^2_t u)(X) + (Au)(X) = 0, \quad X \in G^\pm,$$

$$u(x, 0+) = (\partial_t u)(x, 0+) = 0, \quad x \in S^\pm,$$

$$(\overline{Tu})^\pm(X) = \bar{g}(X), \quad u^\pm_3(X) = f_3(X), \quad X \in \Gamma,$$

where $\bar{v} = (v_1, v_2)$ and the subscript 3 indicates the third component of the corresponding vector-valued function.
   We denote by $\bar{\gamma}^\pm$ and $\gamma^\pm_3$ the trace operators acting on functions $u \in H_{1,p}(S^\pm)$, $p \in \mathbb{C}_0$, according to the formulas

$$\bar{\gamma}^\pm u = \overline{\gamma^\pm u}, \quad \gamma^\pm_3 u = (\gamma^\pm u)_3.$$

$\bar{H}_{1,p}(S^\pm)$ and $H_{1,p}^{(3)}(S^\pm)$ are the subspaces of $H_{1,p}(S^\pm)$ that consist of all u such that $\gamma_3^\pm u = 0$ and $\bar{\gamma}^\pm u = 0$, respectively.

$\bar{H}_{\pm 1/2,p}(\partial S)$ and $H_{\pm 1/2,p}^{(3)}(\partial S)$ are defined just as $H_{\pm 1/2,p}(\partial S)$ but consist of two-component vector-valued functions and scalar functions, respectively. For simplicity, the norms on all these spaces are denoted by $\|\cdot\|_{\pm 1/2,p;\partial S}$.

$\bar{H}_{\pm 1/2,k,\kappa}(\Gamma)$ and $H_{\pm 1/2,k,\kappa}^{(3)}(\Gamma)$, $\kappa > 0$, $k \in \mathbb{R}$, and their norms $\|\cdot\|_{\pm 1/2,k,\kappa;\Gamma}$ are introduced by analogy with the above.

The trace operators in the spaces of originals which correspond to $\bar{\gamma}^\pm$ and $\gamma_3^\pm$ are denoted by the same symbols.

The variational versions of $(\mathrm{DC}_1^\pm)$ consist in finding $u \in H_{1,0,\kappa}(G^\pm)$ that satisfy

$$\int\limits_0^\infty \left\{ a_\pm(u,v) - (B^{1/2}\partial_t u, B^{1/2}\partial_t v)_{0;S^\pm} \right\} dt$$

$$= \pm \int\limits_0^\infty (\bar{g}, \bar{v}^\pm)_{0;\partial S}\, dt \quad \forall v \in C_0^\infty(\bar{G}^\pm),\ \gamma_3^\pm v^\pm = 0, \qquad (7.35)$$

$$\gamma_3^\pm u = f_3.$$

7.7 Theorem. *For any* $f_3 \in H_{1/2,1,\kappa}^{(3)}(\Gamma)$ *and* $\bar{g} \in \bar{H}_{-1/2,1,\kappa}(\Gamma)$, $\kappa > 0$, *problems (7.35) have unique solutions* $u \in H_{1,0,\kappa}(G^\pm)$. *If* $f_3 \in H_{1/2,k,\kappa}^{(3)}(\Gamma)$ *and* $\bar{g} \in \bar{H}_{-1/2,k,\kappa}(\Gamma)$, $k \in \mathbb{R}$, *then* $u \in H_{1,k-1,\kappa}(G^\pm)$ *and*

$$\|u\|_{1,k-1,\kappa;G^\pm} \le c\big(\|f_3\|_{1/2,k,\kappa;\Gamma} + \|\bar{g}\|_{-1/2,k,\kappa;\Gamma}\big). \qquad (7.36)$$

The proof of this assertion is based on the application of the usual procedure, so we omit it.

Let $p \in \mathbb{C}_0$, and let $u \in H_{1,p}(S^\pm)$ be the solutions of the problems

$$p^2(B^{1/2}u, B^{1/2}v)_{0;S^\pm} + a_\pm(u,v) = 0, \quad \forall v \in \mathring{H}_{1,p}(S^\pm),$$

$$\gamma^\pm u = f, \qquad (7.37)$$

where $f \in H_{1/2,p}(\partial S)$. We define operators $\pi_{p,3}^\pm$ on $H_{1/2,p}(\partial S)$ by

$$\pi_{p,3}^\pm f = \{\overline{\mathcal{T}_p^\pm f}, f_3\},$$

where $\mathcal{T}_p^\pm$ are the Poincaré–Steklov operators constructed with respect to the domains $S^\pm$. The operators $\theta_{p,3}^\pm$ are defined on the space $H_{-1/2,p}(\partial S)$ by

$$\theta_{p,3}^\pm = \pi_{p,3}^\pm (\mathcal{T}_p^\pm)^{-1}.$$

7.8 Lemma. *For any number $p \in \mathbb{C}_0$, the operators $\pi_{p,3}^{\pm}$ and $\theta_{p,3}^{\pm}$ are homeomorphisms from the spaces $H_{1/2,p}(\partial S)$ and $H_{-1/2,p}(\partial S)$, respectively, to $\bar{H}_{-1/2,p}(\partial S) \times H_{1/2,p}^{(3)}(\partial S)$, and for any $f \in H_{1/2,p}(\partial S)$ and $g \in H_{-1/2,p}(\partial S)$, $p \in \bar{\mathbb{C}}_\kappa$, $\kappa > 0$,*

$$\|\pi_{p,3}^{\pm} f\|_{\bar{H}_{-1/2,p}(\partial S) \times H_{1/2,p}^{(3)}(\partial S)} \leq c|p| \|f\|_{1/2,p;\partial S}, \tag{7.38}$$

$$\|f\|_{1/2,p;\partial S} \leq c|p| \|\pi_{p,3}^{\pm} f\|_{\bar{H}_{-1/2,p}(\partial S) \times H_{1/2,p}^{(3)}(\partial S)}, \tag{7.39}$$

$$\|\theta_{p,3}^{\pm} g\|_{\bar{H}_{-1/2,p}(\partial S) \times H_{1/2,p}^{(3)}(\partial S)} \leq c|p| \|g\|_{-1/2,p;\partial S}, \tag{7.40}$$

$$\|g\|_{-1/2,p;\partial S} \leq c|p| \|\theta_{p,3}^{\pm} g\|_{\bar{H}_{-1/2,p}(\partial S) \times H_{1/2,p}^{(3)}(\partial S)}. \tag{7.41}$$

Proof. The continuity of $\pi_{p,3}^{\pm}$ and (7.38) follow from Lemma 3.1. Let $u \in H_{1,p}(S^{\pm})$ be the (unique) solutions of the problems

$$p^2(B^{1/2}u, B^{1/2}v)_{0;S^{\pm}} + a_{\pm}(u,v)$$

$$= \pm(\bar{g}, \gamma^{\pm}\bar{v})_{0;\partial S} \quad \forall v \in \bar{H}_{1,p}(S^{\pm}), \tag{7.42}$$

$$\gamma_3^{\pm} u = f_3.$$

By (7.36),

$$\|u\|_{1,p;S^{\pm}} \leq c|p|(\|f_3\|_{1/2,p;\partial S} + \|\bar{g}\|_{-1/2,p;\partial S}). \tag{7.43}$$

The trace theorem and (9.3) imply that (7.39) holds, which proves the statement concerning $\pi_{p,3}^{\pm}$.

By Lemma 3.1 and the properties of $\pi_{p,3}^{\pm}$, the operators $\theta_{p,3}^{\pm}$ are homeomorphisms from $H_{-1/2,p}(\partial S)$ to $\bar{H}_{-1/2,p}(\partial S) \times H_{1/2,p}^{(3)}(\partial S)$. Estimate (7.40) follows from Lemma 3.1. If u is the solution of (7.42), then formulas (7.36), written in terms of Laplace transforms (but with the hat omitted from the transform symbol), and (3.5) imply that (7.41) holds. □

We represent the solutions of problems $(\mathrm{DC}_1^{\pm})$ in the form

$$u(X) = (V\alpha)(X), \quad X \in G^{\pm}, \tag{7.44}$$

or $u = \pi^{\pm} V\alpha$. This leads to the systems of boundary equations

$$\overline{\mathcal{T}^{\pm} V_0 \alpha} = \bar{g}, \quad (V_0 \alpha)_3 = f_3. \tag{7.45}$$

7.9 Theorem. *For any given $f_3 \in H_{1/2,k,\kappa}^{(3)}(\Gamma)$ and $\bar{g} \in \bar{H}_{-1/2,k,\kappa}(\Gamma)$, $\kappa > 0$, $k \in \mathbb{R}$, systems (7.45) have unique solutions $\alpha \in H_{-1/2,k-2,\kappa}(\Gamma)$, in which case the functions u defined by (7.44) belong, respectively, to $H_{1,k-1,\kappa}(G^{\pm})$. If $k \geq 1$, then these functions are the weak solutions of problems $(\mathrm{DC}_1^{\pm})$.*

Proof. In terms of Laplace transforms, (7.45) becomes

$$\overline{T_p^{\pm} V_{p,0} \hat{\alpha}} = \bar{\hat{g}}, \quad (V_{p,0} \hat{\alpha})_3 = \hat{f}_3,$$

or

$$\pi_{p,3}^{\pm} V_{p,0} \hat{\alpha} = \{\bar{\hat{g}}, \hat{g}_3\}. \tag{7.46}$$

By Lemmas 7.8 and 4.1, equation (7.46) is uniquely solvable, and its solution satisfies

$$\|\hat{\alpha}\|_{-1/2,p;\partial S} \le c|p|^2 \|\{\bar{\hat{g}}, \hat{f}_3\}\|_{\bar{H}_{-1/2,p}(\partial S) \times H_{1/2,p}^{(3)}(\partial S)}. \tag{7.47}$$

Taking (7.47) and (7.36) into account, we complete the proof in the usual way. □

We now represent the solutions of our problems in the form

$$u(X) = (W\beta)(X), \quad X \in G^{\pm}, \tag{7.48}$$

or

$$u = \pi^{\pm} W \beta,$$

and arrive at the systems of boundary equations

$$\overline{N\beta} = \bar{g}, \quad (W^{\pm}\beta)_3 = f_3. \tag{7.49}$$

7.10 Theorem. *For any* $f_3 \in H_{1/2,k,\kappa}^{(3)}(\Gamma)$ *and* $\bar{g} \in \bar{H}_{-1/2,k,\kappa}(\Gamma)$, $\kappa > 0$, $k \in \mathbb{R}$, *systems (7.49) have unique solutions* $\beta \in H_{1/2,k-2,\kappa}(\Gamma)$, *in which case the functions* u *defined by (7.48) belong, respectively, to* $H_{1,k-1,\kappa}(G^{\pm})$. *If* $k \ge 1$, *then these functions are the weak solutions of problems* (DC$_1^{\pm}$).

Proof. Going over to Laplace transforms, we bring (7.49) to the form

$$\overline{N_p \hat{\beta}} = \bar{\hat{g}}, \quad (W_p^{\pm} \hat{\beta})_3 = \hat{f}_3,$$

or

$$\theta_{p,3}^{\pm} N_p \hat{\beta} = \{\bar{\hat{g}}, \hat{f}_3\}. \tag{7.50}$$

By Lemma 7.8 and Theorem 4.8, (7.50) has a unique solution $\hat{\beta} \in H_{1/2,p}(\partial S)$, which satisfies

$$\|\hat{\beta}\|_{1/2,p;\partial S} \le c|p|^2 \|\{\bar{\hat{g}}, \hat{f}_3\}\|_{\bar{H}_{-1/2,p}(\partial S) \times H_{1/2,p}^{(3)}(\partial S)}. \tag{7.51}$$

Using (7.51) and (7.36), we now complete the proof according to the standard scheme. □

The classical interior and exterior initial-boundary value problems $(DC_2^\pm)$ consist in finding $u \in C^2(G^\pm) \cap C^1(\bar{G}^\pm)$ such that

$$B(\partial_t^2 u)(X) + (Au)(X) = 0, \quad X \in G^\pm,$$

$$u(x,0+) = (\partial_t u)(x,0+) = 0, \quad x \in S^\pm,$$

$$\bar{u}^\pm(X) = \bar{f}(X), \quad (Tu)_3^\pm(X) = g_3(X), \quad X \in \Gamma,$$

where $\bar{f}$ and g_3 are prescribed on Γ.

We call $u \in H_{1,0,\kappa}(G^\pm)$ weak solutions of $(DC_2^\pm)$ if

$$\int_0^\infty \left\{ a_\pm(u,v) - (B^{1/2}\partial_t u, B^{1/2}\partial_t v)_{0;S^\pm} \right\} dt$$

$$= \pm \int_0^\infty (g_3, v_3^\pm)_{0;\partial S}\, dt \quad \forall v \in C_0^\infty(\bar{G}^\pm), \ \bar{\gamma}^\pm v^\pm = 0, \qquad (7.52)$$

$$\bar{\gamma}^\pm u = \bar{f}.$$

7.11 Theorem. *For any $\bar{f} \in \bar{H}_{1/2,1,\kappa}(\Gamma)$ and $g_3 \in H^{(3)}_{-1/2,1,\kappa}(\Gamma)$, $\kappa > 0$, problems (7.52) have unique solutions $u \in H_{1,0,\kappa}(G^\pm)$. If $\bar{f} \in \bar{H}_{1/2,k,\kappa}(\Gamma)$ and $g_3 \in H^{(3)}_{-1/2,k,\kappa}(\Gamma)$, $k \in \mathbb{R}$, then $u \in H_{1,k-1,\kappa}(G^\pm)$ and*

$$\|u\|_{1,k-1,\kappa;G^\pm} \le c(\|\bar{f}\|_{1/2,k,\kappa;\Gamma} + \|g_3\|_{-1/2,k,\kappa;\Gamma}). \qquad (7.53)$$

Once again, since the proof of this assertion is carried out according to the standard procedure, we omit it.

Let $u \in H_{1,p}(S^\pm)$ be the solutions of (7.37). We define operators $\bar{\pi}_p^\pm$ on $H_{1/2,p}(\partial S)$ and $\bar{\theta}_p^\pm$ on $H_{-1/2,p}(\partial S)$ by

$$\bar{\pi}_p^\pm f = \{\bar{f}, (\mathcal{T}_p^\pm f)_3\},$$

$$\bar{\theta}_p^\pm = \bar{\pi}_p^\pm (\mathcal{T}_p^\pm)^{-1}.$$

7.12 Lemma. *For any number $p \in \mathbb{C}_0$, the operators $\bar{\pi}_p^\pm$ and $\bar{\theta}_p^\pm$ are homeomorphisms from $H_{1/2,p}(\partial S)$ and $H_{-1/2,p}(\partial S)$, respectively, to $\bar{H}_{1/2,p}(\partial S) \times H^{(3)}_{-1/2,p}(\partial S)$, and for any $f \in H_{1/2,p}(\partial S)$ and $g \in H_{-1/2,p}(\partial S)$, $p \in \bar{\mathbb{C}}_\kappa$, $\kappa > 0$,*

$$\|\bar{\pi}_p^\pm f\|_{\bar{H}_{1/2,p}(\partial S) \times H^{(3)}_{-1/2,p}(\partial S)} \le c|p|\|f\|_{1/2,p;\partial S},$$

$$\|f\|_{1/2,p;\partial S} \le c|p|\|\bar{\pi}_p^\pm f\|_{\bar{H}_{1/2,p}(\partial S) \times H^{(3)}_{-1/2,p}(\partial S)},$$

$$\|\bar{\theta}_p^\pm g\|_{\bar{H}_{1/2,p}(\partial S) \times H^{(3)}_{-1/2,p}(\partial S)} \le c|p|\|g\|_{-1/2,p;\partial S},$$

$$\|g\|_{-1/2,p;\partial S} \le c|p|\|\bar{\theta}_p^\pm g\|_{\bar{H}_{1/2,p}(\partial S) \times H^{(3)}_{-1/2,p}(\partial S)}.$$

The proof of this assertion is a repeat of that of Lemma 7.3, with the obvious changes.

We now seek the solutions of $(DC_2^{\pm})$ in the form (7.44). This yields the systems of boundary equations

$$\overline{V_0 \alpha} = \bar{f}, \quad (\mathcal{T}^{\pm} V_0 \alpha)_3 = g_3. \tag{7.54}$$

7.13 Theorem. *For any given* $\bar{f} \in \bar{H}_{1/2,k,\kappa}(\Gamma)$ *and* $g_3 \in H^{(3)}_{-1/2,k,\kappa}(\Gamma)$, $\kappa > 0$, $k \in \mathbb{R}$, *systems* (7.54) *have unique solutions* $\alpha \in H_{-1/2,k-2,\kappa}(\Gamma)$, *in which case the functions* u *defined by* (7.44) *belong, respectively, to the spaces* $H_{1,k-1,\kappa}(G^{\pm})$. *If* $k \geq 1$, *then these functions are the weak solutions of problems* $(DC_2^{\pm})$.

Proof. In terms of Laplace transforms, (7.54) is written as

$$\overline{V_{p,0} \hat{\alpha}} = \hat{\bar{f}}, \quad (\mathcal{T}^{\pm}_p V_{p,0} \hat{\alpha})_3 = \hat{g}_3,$$

or

$$\bar{\pi}^{\pm}_p V_{p,0} \hat{\alpha} = \{\hat{\bar{f}}, \hat{g}_3\}. \tag{7.55}$$

By Lemmas 7.12 and 4.1, (7.55) have unique solutions $\hat{\alpha} \in H_{-1/2,p}(\partial S)$, which satisfy

$$\|\hat{\alpha}\|_{-1/2,p;\partial S} \leq c|p|^2 \|\{\hat{\bar{f}}, \hat{g}_3\}\|_{\bar{H}_{1/2,p}(\partial S) \times H^{(3)}_{-1/2,p}(\partial S)}. \tag{7.56}$$

Estimates (7.56) and (7.53) now enable us to complete the proof by following the usual procedure. $\square$

Representation (7.48) for the solutions $(DC_2^{\pm})$ leads to the systems of boundary equations

$$\overline{W^{\pm} \beta} = \bar{f}, \quad (N\beta)_3 = g_3. \tag{7.57}$$

7.14 Theorem. *For any* $\bar{f} \in \bar{H}_{1/2,k,\kappa}(\Gamma)$ *and* $g_3 \in H^{(3)}_{-1/2,k,\kappa}(\Gamma)$, $\kappa > 0$, $k \in \mathbb{R}$, *systems* (7.57) *have unique solutions* $\beta \in H_{1/2,k-2,\kappa}(\Gamma)$, *in which case the functions* u *defined by* (7.48) *belong, respectively, to* $H_{1,k-1,\kappa}(G^{\pm})$. *If* $k \geq 1$, *then these functions are the weak solutions of problems* $(DC_2^{\pm})$.

Proof. In the Laplace transform domain, (7.57) become

$$\overline{W^{\pm}_p \hat{\beta}} = \hat{\bar{f}}, \quad (N_p \hat{\beta})_3 = \hat{g}_3,$$

or

$$\bar{\theta}^{\pm}_p N_p \hat{\beta} = \{\hat{\bar{f}}, \hat{g}_3\}. \tag{7.58}$$

By Lemma 7.12 and Theorem 4.8, systems (7.58) have unique solutions $\beta \in H_{1/2,p}(\partial S)$, which satisfy

$$\|\hat{\beta}\|_{1/2,p;\partial S} \leq c|p|^2 \|\{\tilde{\hat{f}}, \hat{g}_3\}\|_{\tilde{H}_{1/2,p}(\partial S) \times H^{(3)}_{-1/2,p}(\partial S)}. \tag{7.59}$$

Inequalities (7.59) and (7.53) are now used to complete the proof be means of the standard scheme.   □

7.3 Elastic Boundary Conditions

Let χ be a self-adjoint (3×3)-matrix-valued function defined on ∂S and such that

(i) the operator of multiplication by χ is continuous from $H_{1/2}(\partial S)$ to $H_{-1/2}(\partial S)$;

(ii) there is a $\chi_0 = \text{const} > 0$ such that

$$(\chi f, f)_{0;\partial S} \geq \chi_0 \|f\|^2_{1/2} \quad \forall f \in H_{1/2}(\partial S).$$

It is easy to see that any constant, positive definite (3×3)-matrix satisfies these conditions.

In the classical dynamic problems $(\text{DR}^\pm)$ with elastic (Robin-type) boundary conditions, we seek a function $u \in C^2(G^\pm) \cap C^1(\bar{G}^\pm)$ such that

$$B(\partial_t^2 u)(X) + (Au)(X) = 0, \quad X \in G^\pm,$$
$$u(x, 0+) = (\partial_t u)(x, 0+) = 0, \quad x \in S^\pm,$$
$$(Tu)^\pm(X) \pm \chi(x)u^\pm(X) = g(X), \quad X \in \Gamma, \ x \in \partial S,$$

where g is prescribed on Γ.

The variational versions of $(\text{DR}^\pm)$ consist in finding $u \in H_{1,0,\kappa}(G^\pm)$ such that

$$\int_0^\infty \left[a_\pm(u,v) - (B^{1/2}\partial_t u, B^{1/2}\partial_t v)_{0;S^\pm} + (\chi u, v^\pm)_{0;\partial S} \right] dt$$

$$= \pm \int_0^\infty (g, v^\pm)_{0;\partial S} \, dt \quad \forall v \in C_0^\infty(\bar{G}^\pm). \tag{7.60}$$

7.15 Theorem. *For any $g \in H_{-1/2,1,\kappa}(\Gamma)$, $\kappa > 0$, problems (7.60) have unique solutions $u \in H_{1,0,\kappa}(G^\pm)$. If $g \in H_{-1/2,k,\kappa}(\Gamma)$, $k \in \mathbb{R}$, then $u \in H_{1,k-1,\kappa}(G^\pm)$ and*

$$\|u\|_{1,k-1,\kappa;G^\pm} \leq c\|g\|_{-1/2,k,\kappa;\Gamma}.$$

This assertion is proved just like Theorem 3.3, with the obvious changes.

Let $f \in H_{1/2,p}(\partial S)$, and let $u \in H_{1,p}(S^{\pm})$ be the solutions of the problems

$$p^2(B^{1/2}u, B^{1/2}v)_{0;S^{\pm}} + a_{\pm}(u, v) = 0 \quad \forall v \in \overset{\circ}{H}_{1,p}(S^{\pm}),$$

$$\gamma^{\pm}u = f. \tag{7.61}$$

We define operators $\mathcal{T}_{p,\chi}^{\pm}$ by means of the equality

$$(\mathcal{T}_{p,\chi}^{\pm}f, \varphi)_{0;\partial S}$$

$$= \pm\{a_{\pm}(u, v) + p^2(B^{1/2}u, B^{1/2}, v)_{0;S^{\pm}} + (\chi f, \varphi)_{0;\partial S}\}, \tag{7.62}$$

where $\varphi \in H_{1/2,p}(\partial S)$ is arbitrary and $v \in H_{1,p}(S^{\pm})$ is such that $\gamma^{\pm}v = \varphi$.

7.16 Lemma. *For any $p \in \mathbb{C}_0$, the operators $\mathcal{T}_{p,\chi}^{\pm}$ are homeomorphisms from $H_{1/2,p}(\partial S)$ to $H_{-1/2,p}(\partial S)$, and for any $f \in H_{1/2,p}(\partial S)$, $p \in \bar{\mathbb{C}}_{\kappa}$, $\kappa > 0$,*

$$\|\mathcal{T}_{p,\chi}^{\pm}f\|_{-1/2,p;\partial S} \leq c|p|\|f\|_{1/2,p;\partial S}, \tag{7.63}$$

$$\|f\|_{1/2,p;\partial S} \leq c|p|\|\mathcal{T}_{p,\chi}^{\pm}f\|_{-1/2,p;\partial S}. \tag{7.64}$$

Proof. By (7.62),

$$\mathcal{T}_{p,\chi}^{\pm}f = \mathcal{T}_p^{\pm}f \pm \chi f.$$

Lemma 3.1 and the inequality

$$\|\chi f\|_{-1/2,p;\partial S} \leq c\|f\|_{1/2,p;\partial S}$$

imply that $\mathcal{T}_{p,\chi}^{\pm}$ are continuous from $H_{1/2,p}(\partial S)$ to $H_{-1/2,p}(\partial S)$ and that (7.63) holds.

From (7.62) it also follows that

$$|p|^2\|B^{1/2}u\|_{0;S^{\pm}}^2 + a_{\pm}(u, u) + (\chi f, f)_{0;\partial S}$$

$$= \pm\sigma^{-1} \operatorname{Re}\{\bar{p}(\mathcal{T}_{p,\chi}^{\pm}f, f)_{0;\partial S}\}, \quad \sigma = \operatorname{Re} p;$$

hence,

$$\|u\|_{1,p;S^{\pm}}^2 \leq \pm c\sigma^{-1} \operatorname{Re}\{\bar{p}(\mathcal{T}_{p,\chi}^{\pm}f, f)_{0;\partial S}\}. \tag{7.65}$$

By the trace theorem and (7.65),

$$\|f\|_{1/2,p;\partial S}^2 \leq c|p|(\mathcal{T}_{p,\chi}^{\pm}f, f)_{0;\partial S}|$$

$$\leq c|p|\|\mathcal{T}_{p,\chi}^{\pm}f\|_{-1/2,p;\partial S}\|f\|_{1/2,p;\partial S},$$

which proves (7.64).

If the ranges of $\mathcal{T}^{\pm}_{p,\chi}$ are not dense in $H_{-1/2,p}(\partial S)$, then there is a nonzero $\psi \in H_{1/2,p}(\partial S)$ such that

$$(\mathcal{T}^{\pm}_{p,\chi} f, \psi)_{0;\partial S} = 0 \quad \forall f \in H_{1/2,p}(\partial S). \tag{7.66}$$

We take $f = \psi$ in (7.66) and construct the corresponding solution $u \in H_{1,p}(S^{\pm})$ of (7.61). By (7.65) and (7.66), $u = 0$; therefore, $\psi = \gamma^{\pm} u = 0$. This contradiction proves the lemma. $\qquad\square$

We now represent the solutions of problems $(DR^{\pm})$ in the form

$$u(X) = (V\alpha)(X), \quad X \in G^{\pm}, \tag{7.67}$$

or $u = \pi^{\pm} V\alpha$. This yields the systems of boundary equations

$$\mathcal{T}^{\pm}_{\chi} V_0 \alpha = g, \tag{7.68}$$

where $\mathcal{T}^{\pm}_{\chi}$ are the boundary operators in the spaces of originals constructed from $\mathcal{T}^{\pm}_{p,\chi}$ in the usual way.

7.17 Theorem. *For any $g \in H_{-1/2,k,\kappa}(\Gamma)$, $\kappa > 0$, $k \in \mathbb{R}$, systems (7.68) have unique solutions $\alpha \in H_{-1/2,k-2,\kappa}(\Gamma)$, in which case the functions u defined by (7.67) belong, respectively, to $H_{1,k-1,\kappa}(G^{\pm})$. If $k > 1$, then these functions are the weak solutions of problems $(DR^{\pm})$.*

Proof. Going over to Laplace transforms, for any $p \in \mathbb{C}_{\kappa}$ we obtain the system of boundary equations

$$\mathcal{T}^{\pm}_{p,\chi} V_{p,0} \hat{\alpha} = \hat{g}. \tag{7.69}$$

By Lemmas 4.1 and 7.16, system (7.69) is uniquely solvable for any $\hat{g} \in H_{-1/2,p}(\partial S)$, and its solution $\hat{\alpha} \in H_{-1/2,p}(\partial S)$ satisfies

$$\|\hat{\alpha}\|_{-1/2,p;\partial S} \leq c|p|^2 \|\hat{g}\|_{-1/2,p;\partial S}. \tag{7.70}$$

Estimate (7.70) and Theorem 7.15 now enable us to complete the proof in the usual way. $\qquad\square$

We introduce the operators

$$N^{\pm}_{p,\chi} = \mathcal{T}^{\pm}_{p,\chi} W^{\pm}_p.$$

7.18 Lemma. (i) *For any $p \in \mathbb{C}_0$, the operators $N^{\pm}_{p,\chi}$ are homeomorphisms from $H_{1/2,p}(\partial S)$ to $H_{-1/2,p}(\partial S)$, and for any function $\hat{\beta} \in H_{1/2,p}(\partial S)$, with $p \in \bar{\mathbb{C}}_{\kappa}$, $\kappa > 0$,*

$$\|N^{\pm}_{p,\chi}\hat{\beta}\|_{-1/2,p;\partial S} \leq c|p|^3 \|\hat{\beta}\|_{1/2,p;\partial S},$$

$$\|\hat{\beta}\|_{1/2,p;\partial S} \leq c|p|^3 \|N^{\pm}_{p,\chi}\hat{\beta}\|_{-1/2,p;\partial S}.$$

(ii) *If χ is such that for $p \in \bar{\mathbb{C}}_{\kappa_0}$, $\kappa_0 > 0$,*

$$\|\chi w\|_{-1/2,p;\partial S} \le c\|w\|_{-1/2,p;\partial S} \quad \forall w \in H_{-1/2,p;\partial S}, \tag{7.71}$$

then there is $\kappa^* > 0$ such that for any $p \in \bar{\mathbb{C}}_{\kappa^*}$,

$$\|\hat{\beta}\|_{1/2,p;\partial S} \le c|p|\|N_{p,\chi}^{\pm}\hat{\beta}\|_{-1/2,p;\partial S}. \tag{7.72}$$

Proof. The first statement follows from Lemmas 4.5 and 7.16.

Suppose now that (7.71) holds, and let $\hat{\beta} \in H_{1/2,p}(\partial S)$. By the trace theorem,

$$\|W_p^+\hat{\beta}\|_{1/2,p;\partial S}^2 + \|W_p^-\hat{\beta}\|_{1/2,p;\partial S}^2$$
$$\le c(\|\pi^+ W_p\hat{\beta}\|_{1,p;S^+}^2 + \|\pi^- W_p\hat{\beta}\|_{1,p;S^-}^2)$$
$$\le c\sigma^{-1} \operatorname{Re}\{\bar{p}(N_p\hat{\beta}, \hat{\beta})_{0;\partial S}\}.$$

Clearly,

$$N_{p,\chi}^{\pm}\hat{\beta} = N_p\hat{\beta} \pm \chi W_p^{\pm}\hat{\beta},$$

so,

$$\|W_p^+\hat{\beta}\|_{1/2,p;\partial S}^2 + \|W_p^-\hat{\beta}\|_{1/2,p;\partial S}^2$$
$$\le \sigma^{-1}\{|p|(N_{p,\chi}^{\pm}\hat{\beta}, \hat{\beta})_{0;\partial S}| + |p|(\chi W_p^{\pm}\hat{\beta}, \hat{\beta})_{0;\partial S}|\}.$$

Let $p \in \bar{\mathbb{C}}_{\kappa_0}$. Then, by (7.71),

$$|p|\|\chi\hat{\beta}\|_{-1/2,p;\partial S} \le c|p|\|\hat{\beta}\|_{-1/2,p;\partial S} \le c\|\hat{\beta}\|_{1/2,p;\partial S}$$

and

$$\|W_p^+\hat{\beta}\|_{1/2,p;\partial S}^2 + \|W_p^-\hat{\beta}\|_{1/2,p;\partial S}^2$$
$$\le c\sigma^{-1}|p|(N_{p\chi}^{\pm}\hat{\beta}, \hat{\beta})_{0;\partial S}| + c\sigma^{-1}\|W_p^{\pm}\hat{\beta}\|_{1/2,p;\partial S}\|\hat{\beta}\|_{1/2,p,\partial S};$$

therefore,

$$\|W_p^+\hat{\beta}\|_{1/2,p;\partial S}^2 + \|W_p^-\hat{\beta}\|_{1/2,p;\partial S}^2$$
$$\le c\sigma^{-1}|(N_{p,\chi}^{\pm}\hat{\beta}, \hat{\beta})_{0;\partial S}| + c\sigma^{-2}\|\hat{\beta}\|_{1/2,p;\partial S}^2.$$

The jump formula $\hat{\beta} = W_p^-\hat{\beta} - W_p^+\hat{\beta}$ implies that

$$\|\hat{\beta}\|_{1/2,p;\partial S}^2 \le c(\|W_p^+\hat{\beta}\|_{1/2,p;\partial S}^2 + \|W_p^-\hat{\beta}\|_{1/2,p;\partial S}^2)$$
$$\le c\sigma^{-1}\|N_{p,\chi}^{\pm}\hat{\beta}\|_{-1/2,p;\partial S}\|\hat{\beta}\|_{1/2,p;\partial S} + c\sigma^{-2}\|\hat{\beta}\|_{1/2,p;\partial S}^2. \tag{7.73}$$

By (7.73), there is $\kappa^* > 0$ such that for $\sigma \geq \kappa^*$,

$$\|\hat{\beta}\|^2_{1/2,p;\partial S} \leq c\sigma^{-1}\|N^\pm_{p,\chi}\hat{\beta}\|_{-1/2,p;\partial S}\|\hat{\beta}\|_{1/2,p;\partial S},$$

which proves (7.72). □

We now seek the solutions of problems $(DR^\pm)$ in the form

$$u(X) = (W\beta)(X), \quad X \in G^\pm, \tag{7.74}$$

or

$$u = \pi^\pm W\beta.$$

This leads to the systems of boundary equations

$$N^\pm_\chi \beta = g, \tag{7.75}$$

where $N^\pm_\chi$ are the boundary operators in the spaces of originals which correspond to $N^\pm_{p,\chi}$.

7.19 Theorem. *For any $g \in H_{-1/2,k,\kappa}(\Gamma)$, $\kappa > 0$, $k \in \mathbb{R}$, systems (7.75) have unique solutions $\beta \in H_{1/2,k-3,\kappa}(\Gamma)$. If χ satisfies (7.71), then there is $\kappa^* > 0$ such that $\beta \in H_{1/2,k-1,\kappa}(\Gamma)$ for $\kappa \geq \kappa^*$, in which case the functions u defined by (7.74) belong, respectively, to $H_{1,k-1,\kappa}(\Gamma)$. If $k \geq 1$, then these functions are the weak solutions of problems $(DR^\pm)$.*

Proof. In the Laplace transform domain, (7.75) becomes

$$N^\pm_{p,\chi}\hat{\beta} = \hat{g}.$$

The assertion now follows from Lemma 7.18 and Theorem 7.15. □

Boundary Integral Equations for Plates on a Generalized Elastic Foundation

8.1 Formulation and Solvability of the Problems

In the case of bending of a plate that lies on a generalized elastic foundation, the equations of motion (1.4) are replaced by

$$\partial_j t_{ij} - (Kv)_i + f_i = \rho \partial_t^2 v_i, \quad i = 1, 2, 3,$$

where $K = (k_{ij})$ is a constant, symmetric, positive definite (3×3)-matrix. The averaging procedure described in §1.1 leads to

$$B(\partial_t^2 u)(X) + (Au)(X) + (\mathcal{K}u)(X) = q(X),$$
$$X \in G = S \times (0, \infty); \qquad (8.1)$$

here

$$\mathcal{K} = \begin{pmatrix} \bar{\mathcal{K}} & 0 \\ 0 & k_{33} \end{pmatrix}, \quad \bar{\mathcal{K}} = h^2 \begin{pmatrix} k_{11} & k_{12} \\ k_{21} & k_{22} \end{pmatrix}.$$

Clearly, $\bar{\mathcal{K}}$ and $\mathcal{K}$ are positive definite matrices.

For simplicity, we restrict our attention to only two types of boundary conditions; the corresponding theory can easily be developed for all the initial-boundary value problems investigated earlier.

The variational problems $(DD_\mathcal{K}^\pm)$ with Dirichlet boundary conditions and zero initial conditions for the homogeneous equation of motion consist in finding $u \in H_{1,0,\kappa}(G^\pm)$ such that

$$\int_0^\infty \left[a_\pm(u, v) + (\mathcal{K}u, v)_{0;S^\pm} - (B^{1/2}\partial_t u, B^{1/2}\partial_t v)_{0;S^\pm} \right] dt = 0$$

$$\forall v \in C_0^\infty(\bar{G}^\pm), \quad v^\pm = 0, \qquad (8.2)$$
$$\gamma^\pm u = f,$$

where f is prescribed on Γ.

Obviously, in terms of Laplace transforms, (8.2) turns into the problem of seeking the (weak) solutions $\hat{u} \in H_{1,p}(S^{\pm})$ of problems $(D_{\mathcal{K},p}^{\pm})$

$$p^2 B\hat{u}(x,p) + (A\hat{u})(x,p) + \mathcal{K}\hat{u}(x,p) = 0, \quad x \in S^{\pm},$$

$$\gamma^{\pm}\hat{u}(x,p) = \hat{f}(x,p), \quad x \in \partial S.$$

8.1 Lemma. *For any $\hat{f} \in H_{1/2,p}(\partial S)$, $p \in \bar{\mathbb{C}}_{\kappa}$, $\kappa > 0$, problems $(D_{\mathcal{K},p}^{\pm})$ have unique solutions $\hat{u} \in H_{1,p}(S^{\pm})$ and*

$$\|\hat{u}\|_{1,p;S^{\pm}} \leq c|p|\|\hat{f}\|_{1/2,p;\partial S}.$$

This assertion is proved just like Theorems 2.1 and 2.2.

8.2 Theorem. *For any $f \in H_{1/2,1,\kappa}(\Gamma)$, $\kappa > 0$, problems (8.2) have unique solutions $u \in H_{1,0,\kappa}(G^{\pm})$. If $f \in H_{1/2,k,\kappa}(\Gamma)$, $k \in \mathbb{R}$, then $u \in H_{1,k-1,\kappa}(\Gamma^{\pm})$ and*

$$\|u\|_{1,k-1,\kappa;G^{\pm}} \leq c\|f\|_{1/2,k,\kappa;\Gamma}.$$

The proof of this statement is similar to that of Theorem 2.3.

In problems $(DN_{\mathcal{K}}^{\pm})$ for the homogeneous equation of motion with Neumann boundary conditions and zero initial conditions, we want to find $u \in H_{1,0,\kappa}(G^{\pm})$ such that

$$\int_0^{\infty} \left[a_{\pm}(u,v) + (\mathcal{K}u,v)_{0;S^{\pm}} - (B^{1/2}\partial_t u, B^{1/2}\partial_t v)_{0;S^{\pm}} \right] dt$$

$$= \pm \int_0^{\infty} (g, v^{\pm})_{0;\partial S} \, dt \quad \forall v \in C_0^{\infty}(\bar{G}^{\pm}), \qquad (8.3)$$

where g is prescribed on Γ.

When the Laplace transformation is applied to (8.3), we arrive at problems $(N_{\mathcal{K},p}^{\pm})$, which consist in finding (weak) solutions $\hat{u} \in H_{1,p}(S^{\pm})$ such that

$$p^2 B\hat{u}(x,p) + (A\hat{u})(x,p) + \mathcal{K}\hat{u}(x,p) = 0, \quad x \in S^{\pm},$$

$$(T\hat{u})^{\pm}(x,p) = \hat{g}(x,p), \quad x \in \partial S.$$

The next two assertions are stated without proof because they are fully analogous to their corresponding counterparts in the case $\mathcal{K} = 0$.

8.3 Lemma. *For any $\hat{g} \in H_{-1/2,p}(\partial S)$, $p \in \bar{\mathbb{C}}_{\kappa}$, $\kappa > 0$, problems $(N_{\mathcal{K},p}^{\pm})$ have unique solutions $\hat{u} \in H_{1,p}(S^{\pm})$ and*

$$\|\hat{u}\|_{1,p;S^{\pm}} \leq c|p|\|\hat{g}\|_{-1/2,p;\partial S}.$$

8.4 Theorem. *For any* $g \in H_{-1/2,1,\kappa}(\Gamma)$, $\kappa > 0$, *problems* (8.3) *have unique solutions* $u \in H_{1,0,\kappa}(G^{\pm})$. *If* $g \in H_{-1/2,k,\kappa}(\Gamma)$, $k \in \mathbb{R}$, *then* $u \in H_{1,k-1,\kappa}(G^{\pm})$ *and*

$$\|u\|_{1,k-1,\kappa;G^{\pm}} \leq c\|g\|_{-1/2,k,\kappa;\Gamma}.$$

For simplicity, in what follows we consider the case when

$$\mathcal{K} = \mathrm{diag}\{h^2 k_0, h^2 k_0, k_3\}, \quad k_0, \; k_3 = \mathrm{const} > 0;$$

however, all the results obtained below remain valid for a general case constant, symmetric, and positive definite matrix $\mathcal{K}$.

8.2 A Matrix of Fundamental Solutions

The construction of single-layer and double-layer plate potentials requires the knowledge of a matrix of fundamental solutions $D_{\mathcal{K}}(x,t)$ for equation (8.1), or of its Laplace transform $\hat{D}_{\mathcal{K}}(x,p)$. The latter is a (3×3)-matrix such that

$$p^2 B\hat{D}_{\mathcal{K}}(x,p) + (A\hat{D}_{\mathcal{K}})(x,p) + \mathcal{K}\hat{D}_{\mathcal{K}}(x,p) = \delta(x)I, \quad x \in \mathbb{R}^2, \qquad (8.4)$$

where δ is the Dirac delta distribution and I is the identity (3×3)-matrix. Going over to Fourier transforms in (8.4), we arrive at

$$p^2 B\tilde{D}_{\mathcal{K}}(\xi,p) + (A(\xi)\tilde{D}_{\mathcal{K}})(\xi,p) + \mathcal{K}\tilde{D}_{\mathcal{K}}(\xi,p) = I,$$

or

$$\Psi_{\mathcal{K}}(\xi,p)\tilde{D}_{\mathcal{K}}(\xi,p) = I, \qquad (8.5)$$

where the entries of the (3×3)-matrix $\Psi_{\mathcal{K}}(\xi,p)$ are

$$\Psi_{\mathcal{K},\alpha\beta}(\xi,p) = h^2(\lambda + \mu)\xi_\alpha\xi_\beta$$
$$+ \delta_{\alpha\beta}\big[h^2(k_0 + \rho p^2) + \mu + h^2\mu|\xi|^2\big],$$
$$\Psi_{\mathcal{K},33}(\xi,p) = k_3 + \rho p^2 + \mu|\xi|^2,$$
$$\Psi_{\mathcal{K},\alpha 3}(\xi,p) = -\Psi_{\mathcal{K},3\alpha}(\xi,p) = -i\mu\xi_\alpha. \qquad (8.6)$$

To obtain the inverse matrix $\Psi_{\mathcal{K}}^{-1}(\xi,p)$, we first need to compute $\det \Psi_{\mathcal{K}}(\xi,p)$. Direct calculation shows that $\det \Psi_{\mathcal{K}}(\xi,p)$ is a function of $|\xi|^2$ alone; in other words, it is rotation-invariant in $\mathbb{R}^2$. Therefore, we may take

$$\xi_1 = |\xi|, \quad \xi_2 = 0$$

to find that

$$\det \Psi_{\mathcal{K}}(\xi, p) = R_{\mathcal{K}}(|\xi|^2, p)$$
$$= \left[h^2 \mu |\xi|^2 + h^2(k_0 + \rho p^2) + \mu \right]$$
$$\times \left\{ h^2 \mu (\lambda + \mu) |\xi|^4 \right.$$
$$+ h^2 \left[(k_3 + \rho p^2)(\lambda + 2\mu) + (k_0 + \rho p^2)\mu \right] |\xi|^2$$
$$\left. + (k_3 + \rho p^2) \left[h^2 (k_0 + \rho p^2) + \mu \right] \right\}.$$

The roots $t_j = -\chi_j^2$, $j = 1, 2, 3$, of the equation $R_{\mathcal{K}}(t, p) = 0$ are

$$\chi_{1,2}^2 = \left(2h\mu(\lambda + 2\mu) \right)^{-1}$$
$$\times \left\{ h \left[(k_3 + \rho p^2)(\lambda + 2\mu) + (k_0 + \rho p^2)\mu \right] \right.$$
$$\pm \left[h^2 \left((k_3 + \rho p^2)(\lambda + 2\mu) - (k_0 + \rho p^2)\mu \right)^2 \right.$$
$$\left. \left. - 4(k_3 + \rho p^2)\mu^2(\lambda + 2\mu) \right]^{1/2} \right\},$$
$$\chi_3^2 = (h^2 \mu)^{-1} \left[h^2(k_0 + \rho p^2) + \mu \right].$$

We choose χ_j so that $\operatorname{Re} \chi_j \geq 0$, $j = 1, 2, 3$.

8.5 Lemma. (i) *The equation $R_{\mathcal{K}}(t, p) = 0$ does not have a triple root for any $p \in \mathbb{C}_0$.*

(ii) *The equation $R_{\mathcal{K}}(t, p) = 0$ has a double root only for finitely many values of $p \in \mathbb{C}_0$.*

(iii) $\operatorname{Re} \chi_j > 0$, $j = 1, 2, 3$.

Proof. (i) It is not difficult to see that $\chi_1^2 = \chi_2^2 = \chi_3^2$ only if

$$k_0 + \rho p^2 = -\frac{2\lambda\mu(\lambda + 2\mu)}{2h^2(\lambda + \mu)^2}$$

or

$$k_0 + \rho p^2 = -\frac{\mu}{h^2},$$

and that either of these equalities leads immediately to a contradiction.

(ii) The validity of the second assertion is obvious.

(iii) First, we remark that $\chi_i \neq 0$. True, if $\chi_i = 0$ for some $i \in \{1, 2, 3\}$, then

$$\rho p^2 = -\left(\frac{\mu}{h^2} + k_0 \right),$$

which contradicts the fact that $p \in \mathbb{C}_0$.

If $\operatorname{Re} \chi_i = 0$ for some i, then $\chi_i^2 < 0$; hence, $\tilde{t} = |\tilde{\xi}|^2 = -\chi_i^2$ is a positive root of the equation $R_{\mathcal{K}}(t, p) = 0$. We take $\tilde{\xi} = (|\tilde{\xi}|, 0)$ and consider a nonzero solution $g = (g_1, g_2, g_3)^{\mathrm{T}}$ of the system of linear algebraic equations $\Psi_K(\tilde{\xi}, p)g = 0$. Multiplying this equality by g^*, we obtain

$$
\begin{aligned}
\big[h^2(\rho p^2 + k_0) + \mu\big]&\big(|g_1|^2 + |g_2|^2\big) \\
&+ (\rho p^2 + k_3 + \mu|\tilde{\xi}|^2)|g_3|^2 + h^2|\tilde{\xi}|^2\big[(\lambda + 2\mu)|g_1|^2 + \mu|g_2|^2\big] \\
&\hspace{6cm} - 2\mu|\tilde{\xi}|\,\mathrm{Re}(g_3^* i g_1) = 0;
\end{aligned}
$$

therefore, $p^2 \in \mathbb{R}$ and $p^2 > 0$. Since

$$
2\mu|\tilde{\xi}|\,\mathrm{Re}(g_3^* i g_1) \geq -\mu\big(|g_1|^2 + |\tilde{\xi}|^2|g_3|^2\big),
$$

we obtain

$$
\begin{aligned}
(\rho p^2 + k_0)h^2\big(|g_1|^2 + |g_2|^2\big) &+ (\rho p^2 + k_3)|g_3|^2 \\
&+ h^2|\tilde{\xi}|^2\big[(\lambda + 2\mu)|g_1|^2 + \mu|g_2|^2\big] \leq 0.
\end{aligned}
$$

This contradiction completes the proof. $\square$

If we write

$$
\tilde{\Psi}_K(\xi, p) = \big[\det \Psi_K(\xi, p)\big]^{-1},
$$

then, by (8.5) and (8.6),

$$
\begin{aligned}
\tilde{D}_{K,11}(\xi, p) = \big[&h^2\mu^2|\xi|^4 + h^2\mu(\lambda + \mu)|\xi|^2\xi_2^2 \\
&+ \mu h^2(k_0 + k_3 + 2\rho p^2)|\xi|^2 \\
&+ \mu^2\xi_1^2 + h^2(\lambda + \mu)(k_3 + \rho p^2)\xi_2^2 \\
&+ \big(h^2(k_0 + p^2) + \mu\big)(k_3 + \rho p^2)\big]\tilde{\Psi}_K(\xi, p), \\[4pt]
\tilde{D}_{K,22}(\xi, p) = \big[&h^2\mu^2|\xi|^4 + h^2\mu(\lambda + \mu)|\xi|^2\xi_1^2 \\
&+ \mu h^2(k_0 + k_3 + 2\rho p^2)|\xi|^2 \\
&+ \mu^2\xi_2^2 + h^2(\lambda + \mu)(k_3 + \rho p^2)\xi_1^2 \\
&+ \big(h^2(k_0 + \rho p^2) + \mu\big)(k_3 + \rho p^2)\big]\tilde{\Psi}_K(\xi, p), \\[4pt]
\tilde{D}_{K,33}(\xi, p) = \big[&h^4\mu(\lambda + \mu)|\xi|^4 + h^2(\lambda + 3\mu)\big(h^2(k_0 + \rho p^2) + \mu\big)|\xi|^2 \\
&+ \big(h^2(k_0 + \rho p^2) + \mu\big)^2\big]\tilde{\Psi}_K(\xi, p), \\[4pt]
\tilde{D}_{K,12}(\xi, p) &= \tilde{D}_{K,21}(\xi, p) \\
&= -\xi_1\xi_2\big[h^2\mu(\lambda + \mu)|\xi|^2 \\
&\hspace{1.5cm} + h^2(\lambda + \mu)(k_3 + \rho p^2) - \mu^2\big]\tilde{\Psi}_K(\xi, p), \\[4pt]
\tilde{D}_{K,\alpha 3}(\xi, p) &= -\tilde{D}_{3\alpha,K}(\xi, p) \\
&= i\mu\xi_\alpha\big[h^2\mu|\xi|^2 + h^2(k_0 + \rho p^2) + \mu\big]\tilde{\Psi}_K(\xi, p).
\end{aligned}
$$

Consequently, applying the inverse Laplace transformation, we find that the entries of $\hat{D}_K(x, p)$ are

$$\hat{D}_{K,11}(x, p) = [h^2\mu^2\Delta^2 + h^2\mu(\lambda + \mu)\Delta\partial_2^2$$
$$- \mu h^2(k_0 + k_3 + 2\rho p^2)\Delta$$
$$- \mu^2\partial_1^2 - h^2(\lambda + \mu)(k_3 + \rho p^2)\partial_2^2$$
$$+ (h^2(k_0 + \rho p^2) + \mu)k_3]\Psi_K(x, p),$$

$$\hat{D}_{K,22}(x, p) = [h^2\mu^2\Delta^2 + h^2\mu(\lambda + \mu)\Delta\partial_1^2$$
$$- h^2\mu(k_0 + k_3 + 2\rho p^2)\Delta$$
$$- \mu^2\partial_2^2 - h^2(\lambda + \mu)(k_3 + \rho p^2)\partial_1^2$$
$$+ (h^2(k_0 + \rho p^2) + \mu)(k_3 + \rho p^2)]\Psi_K(x, p), \qquad (8.7)$$

$$\hat{D}_{K,33}(x, p) = [h^4\mu(\lambda + 2\mu)\Delta^2 - h^2(\lambda + 3\mu)(h^2(k_0 + \rho p^2) + \mu)\Delta$$
$$+ (h^2(k_0 + \rho p^2) + \mu)^2]\Psi_K(x, p),$$

$$\hat{D}_{K,12}(x, p) = \hat{D}_{K,21}(x, p)$$
$$= [-h^2\mu(\lambda + \mu)\Delta$$
$$+ h^2(\lambda + \mu)(k_3 + \rho p^2) - \mu^2]\partial_1\partial_2\Psi_K(x, p),$$

$$\hat{D}_{K,\alpha 3}(x, p) = -\hat{D}_{K,3\alpha}(x, p)$$
$$= \mu[h^2\mu\Delta - h^2(k_0 + \rho p^2) - \mu]\partial_\alpha\Psi_K(x, p),$$

where

$$\Psi_K(x, p) = (4\pi^2)^{-1}\int_{\mathbb{R}^2} e^{-i(x,\xi)}\tilde{\Psi}_K(\xi, p)d\xi$$

is the inverse Fourier transform of $\tilde{\Psi}_K(\xi, p)$.

There are two possible cases. If the equation $R_K(t, p) = 0$ has distinct roots, then we easily convince ourselves that

$$\Psi_K(x, p) = [2\pi h^4\mu^2(\lambda + 2\mu)]^{-1}c_i K_0(\chi_i|x|),$$

where

$$c_1 = [(\chi_1^2 - \chi_3^2)(\chi_1^2 - \chi_2^2)]^{-1},$$
$$c_2 = [(\chi_1^2 - \chi_2^2)(\chi_2^2 - \chi_3^2)]^{-1},$$
$$c_3 = [(\chi_1^2 - \chi_3^2)(\chi_2^2 - \chi_3^2)]^{-1}.$$

If, on the other hand, $R_K(t, p) = 0$ has a double root, say,

$$\chi_1^2 = \chi_2^2 \neq \chi_3^2,$$

then

$$\Psi_K(x,p) = \left[2\pi h^4 \mu^2 (\lambda + 2\mu)\right]^{-1}$$
$$\times \left[\tilde{c}_1 K_0(\chi_1|x|) + \tilde{c}_3 K_0(\chi_3|x|)\right.$$
$$\left. + \tilde{c}_2(2\chi_1)^{-1}|x|K_1(\chi_1|x|)\right],$$

where

$$\tilde{c}_1 = -\tilde{c}_3 = -(\chi_3^2 - \chi_1^2), \quad \tilde{c}_2 = (\chi_3^2 - \chi_1^2)^{-1}.$$

8.6 Lemma. *For any $p \in \mathbb{C}_0$, the function $\Psi_K(x,p)$ can be represented near $x = 0$ in the form*

$$\Psi_K(x,p) = -\left[128\pi h^4 \mu^2 (\lambda + 2\mu)\right]^{-1}|x|^4 \ln|x|$$
$$+ O(|x|^6 \ln|x|) + \Psi_{K,0}(x,p), \qquad (8.8)$$

where $\Psi_{K,0}(x,p)$ is infinitely differentiable. In addition, $\Psi_K(x,p) \to 0$ exponentially as $|x| \to \infty$.

This assertion is proved just like Lemma 1.2.

8.7 Corollary. *For any $p \in \mathbb{C}_0$, the elements of the matrix of fundamental solutions $\hat{D}_K(x,p)$ can be represented in the neighborhood of $x = 0$ in the form*

$$\hat{D}_{K,\alpha\beta}(x,p) = \left[4\pi\mu(\lambda + 2\mu)h^2\right]^{-1}$$
$$\times \left[(\lambda + \mu)x_\alpha x_\beta |x|^{-2} - (\lambda + 3\mu)\delta_{\alpha\beta} \ln|x|\right]$$
$$+ O(|x|^2 \ln|x|) + \hat{D}_{K,0,\alpha\beta}(x,p),$$

$$\hat{D}_{K,33}(x,p) = -(2\pi h^2 \mu)^{-1} \ln|x| \qquad (8.9)$$
$$+ O(|x|^2 \ln|x|) + \hat{D}_{K,0,33}(x,p),$$

$$\hat{D}_{K,\alpha 3}(x,p) = -\hat{D}_{3\alpha K}(x,p) = -\left[4\pi h^2 (\lambda + 2\mu)\right]^{-1} x_\alpha \ln|x|$$
$$+ O(|x|^2 \ln|x|) + \hat{D}_{K,0,\alpha 3}(x,p),$$

where $\hat{D}_{K,0,ij}(x,p)$ are infinitely differentiable functions.

The proof of this assertion follows from (8.7) and (8.8).

8.8 Remark. Equalities (8.9) show that for any $p \in \mathbb{C}_0$, the asymptotic behavior of $\hat{D}_K(x,p)$ near $x = 0$ coincides with that of the matrix of fundamental solutions $D(x)$ in the equilibrium case [10].

8.3 Properties of the Boundary Operators

Let $u \in H_{1,p}(S^\pm)$ be the (unique) solutions of problems $(D_{\mathcal{K},p}^\pm)$ with boundary data $f \in H_{1/2,p}(\partial S)$. We define Poincaré–Steklov operators $\mathcal{T}_{\mathcal{K},p}^\pm$ on $H_{1/2,p}(\partial S)$ through the equality

$$(\mathcal{T}_{\mathcal{K},p}^\pm f, \varphi)_{0;\partial S} = \pm\big[p^2(B^{1/2}u, B^{1/2}w)_{0,S^\pm} + a_{\mathcal{K},\pm}(u,w)\big],$$

where $\varphi \in H_{1/2,p}(\partial S)$, w is any element of $H_{1,p}(S^\pm)$ such that $\gamma^\pm w = \varphi$, and $a_{\mathcal{K},\pm}$ are the bilinear forms defined by

$$a_{\mathcal{K},\pm}(u,w) = a_\pm(u,w) + (\mathcal{K}u, w)_{0,S^\pm}.$$

8.9 Lemma. *For any $p \in \mathbb{C}_0$, the operators $\mathcal{T}_{\mathcal{K},p}^\pm$ are homeomorphisms from $H_{1/2,p}(\partial S)$ to $H_{-1/2,p}(\partial S)$, and for any $f \in H_{1/2,p}(\partial S)$, $p \in \bar{\mathbb{C}}_\kappa$, $\kappa > 0$,*

$$\|\mathcal{T}_{\mathcal{K},p}^\pm f\|_{-1/2,p;\partial S} \le c|p|\|f\|_{1/2,p;\partial S},$$

$$\|f\|_{1/2,p;\partial S} \le c|p|\|\mathcal{T}_{\mathcal{K},p}^\pm f\|_{-1/2,p;\partial S}.$$

The proof of this assertion is identical to that of Lemma 3.1, with the appropriate changes.

At this stage, we introduce the operators $\hat{\mathcal{T}}_\mathcal{K}^\pm$, $(\hat{\mathcal{T}}_\mathcal{K}^\pm)^{-1}$ and $\mathcal{T}_\mathcal{K}^\pm$, $(\mathcal{T}_\mathcal{K}^\pm)^{-1}$ in the usual way.

8.10 Theorem. *For any $\kappa > 0$ and $k \in \mathbb{R}$, the operators $\mathcal{T}_\mathcal{K}^\pm$ are continuous and injective from $H_{1/2,k,\kappa}(\Gamma)$ to $H_{-1/2,k-1,\kappa}(\Gamma)$, and their ranges are dense in $H_{-1/2,k-1,\kappa}(\Gamma)$. The inverses $(\mathcal{T}^\pm)^{-1}$, extended by continuity from the ranges of $\mathcal{T}^\pm$ to $H_{-1/2,k,\kappa}(\Gamma)$, are continuous and injective from $H_{-1/2,k,\kappa}(\Gamma)$ to $H_{1/2,k-1,\kappa}(\Gamma)$ for any $k \in \mathbb{R}$, and their ranges are dense in $H_{1/2,k-1,\kappa}(\Gamma)$.*

The assertion follows from Lemma 8.9 and the arguments used in the proof of Theorem 3.2.

The time-dependent (retarded) single-layer and double-layer potentials are now defined by

$$(V_\mathcal{K}\alpha)(X) = \int\limits_{-\infty}^{\infty} \int\limits_{\partial S} D_\mathcal{K}(x-y, t-\tau)\alpha(y,\tau)\,ds_y\,d\tau,$$

and

$$(W_\mathcal{K}\beta)(X) = \int\limits_{-\infty}^{\infty} \int\limits_{\partial S} (\beta(y,\tau), (T_y D_\mathcal{K}^{(j)})(y-x, t-\tau))e_j\,ds_y\,d\tau.$$

In view of (8.8), the boundary properties of $V_\mathcal{K}\alpha$ and $W_\mathcal{K}\beta$ coincide with those of the corresponding plate potentials in the case $K = 0$.

The boundary operators $V_{\mathcal{K},0}$, $W_{\mathcal{K}}^{\pm}$, and

$$N_{\mathcal{K}} = \mathcal{T}_{\mathcal{K}}^{+} W_{\mathcal{K}}^{+} = \mathcal{T}_{\mathcal{K}}^{-} W_{\mathcal{K}}^{-}$$

associated with the above potentials are introduced according to the scheme used in §4.1. Their properties are stated in the next three assertions, whose proofs are essentially identical to those of Theorem 4.3, 4.6, and 4.8.

8.11 Theorem. *For any $\kappa > 0$ and $k \in \mathbb{R}$, the operator $V_{\mathcal{K},0}$ is continuous and injective from $H_{-1/2,k,\kappa}(\Gamma)$ to $H_{1/2,k-1,\kappa}(\Gamma)$, and its range is dense in $H_{1/2,k-1,\kappa}(\Gamma)$. The inverse $(V_{\mathcal{K},0})^{-1}$, extended by continuity from the range of $V_{\mathcal{K},0}$ to $H_{1/2,k,\kappa}(\Gamma)$, is continuous and injective from $H_{1/2,k,\kappa}(\Gamma)$ to $H_{-1/2,k-1,\kappa}(\Gamma)$ for any $k \in \mathbb{R}$, and its range is dense in $H_{-1/2,k-1,\kappa}(\Gamma)$. In addition, for any $\alpha \in H_{-1/2,k,\kappa}(\Gamma)$,*

$$\|\pi^{+} V_{\mathcal{K}} \alpha\|_{1,k-1,\kappa;G^{+}} + \|\pi^{-} V_{\mathcal{K}} \alpha\|_{1,k-1,\kappa;G^{-}} \leq c\|\alpha\|_{-1/2,k,\kappa;\Gamma}.$$

8.12 Theorem. *For any $\kappa > 0$ and $k \in \mathbb{R}$, the operators $W_{\mathcal{K}}^{\pm}$ are continuous and injective from $H_{1/2,k,\kappa}(\Gamma)$ to $H_{1/2,k-2,\kappa}(\Gamma)$, and their ranges are dense in $H_{1/2,k-2,\kappa}(\Gamma)$. The inverses $(W_{\mathcal{K}}^{\pm})^{-1}$, extended by continuity from the ranges of $W_{\mathcal{K}}^{\pm}$, respectively, to $H_{1/2,k,\kappa}(\Gamma)$ are continuous and injective from $H_{1/2,k,\kappa}(\Gamma)$ to $H_{1/2,k-2,\kappa}(\Gamma)$ for any $k \in \mathbb{R}$, and their ranges are dense in $H_{1/2,k-2,\kappa}(\Gamma)$. In addition, for any $\beta \in H_{1/2,k,\kappa}(\Gamma)$,*

$$\|\pi^{+} W_{\mathcal{K}} \beta\|_{1,k-2,\kappa;G^{+}} + \|\pi^{-} W_{\mathcal{K}} \beta\|_{1,k-2,\kappa;G^{-}} \leq c\|\beta\|_{1/2,k,\kappa;\Gamma}.$$

8.13 Theorem. *For any $\kappa > 0$ and $k \in \mathbb{R}$, the operator $N_{\mathcal{K}}$ is continuous and injective from $H_{1/2,k,\kappa}(\Gamma)$ to $H_{-1/2,k-3,\kappa}(\Gamma)$, and its range is dense in $H_{-1/2,k-3,\kappa}(\Gamma)$. The inverse $N_{\mathcal{K}}^{-1}$, extended by continuity from the range of $N_{\mathcal{K}}$ to $H_{-1/2,k,\kappa}(\Gamma)$, is continuous and injective from $H_{-1/2,k,\kappa}(\Gamma)$ to $H_{1/2,k-1,\kappa}(\Gamma)$ for any $k \in \mathbb{R}$, and its range is dense in $H_{1/2,k-1,\kappa}(\Gamma)$.*

8.4 Solvability of the Boundary Equations

We seek the solutions u of $(\mathrm{DD}_{\mathcal{K}}^{\pm})$ in the form

$$u = \pi^{\pm} V_{\mathcal{K}} \alpha \tag{8.10}$$

or

$$u = \pi^{\pm} W_{\mathcal{K}} \beta. \tag{8.11}$$

Representations (8.10) and (8.11) yield, respectively, the systems of boundary equations

$$V_{\mathcal{K},0} \alpha = f \tag{8.12}$$

and

$$W_{\mathcal{K}}^{\pm} \beta = f. \tag{8.13}$$

The next assertion follows from Theorems 8.11 and 8.12.

8.14 Theorem. *For any* $f \in H_{1/2,k,\kappa}(\Gamma)$, $k \in \mathbb{R}$, $\kappa > 0$, *systems* (8.12) *and* (8.13) *have unique solutions* $\alpha \in H_{-1/2,k-1,\kappa}(\Gamma)$ *and* $\beta \in H_{1/2,k-2,\kappa}(\Gamma)$, *in which case the functions* u *defined by* (8.10) *or* (8.11) *belong to* $H_{1,k-1,\kappa}(G^{\pm})$. *If* $k \geq 1$, *then these functions are the solutions of problems* $(\mathrm{DD}_{\mathcal{K}}^{\pm})$, *respectively.*

We now seek the solutions of $(\mathrm{DN}_{\mathcal{K}}^{\pm})$ in the form (8.10) and (8.11). These representations lead to the systems of boundary equations

$$\mathcal{T}_{\mathcal{K}}^{\pm}(V_{\mathcal{K},0}\alpha) = g \tag{8.14}$$

and

$$N_{\mathcal{K}}\beta = g. \tag{8.15}$$

8.15 Theorem. *For any* $g \in H_{-1/2,k,\kappa}(\Gamma)$, $k \in \mathbb{R}$, $\kappa > 0$, *systems* (8.14) *and* (8.15) *have unique solutions* $\alpha \in H_{-1/2,k-2,\kappa}(\Gamma)$ *and* $\beta \in H_{1/2,k-1,\kappa}(\Gamma)$, *in which case the functions* u *defined by* (8.10) *or* (8.11) *belong to the space* $H_{1,k-1,\kappa}(G^{\pm})$. *If* $k \geq 1$, *then these functions are the solutions of problems* $(\mathrm{DN}_{\mathcal{K}}^{\pm})$, *respectively.*

The assertion follows from Theorems 8.10, 8.11, and 8.13.

8.16. Remark. As shown in Chapter 9, the general case with a nonhomogeneous equation of motion and nonhomogeneous initial conditions can be reduced to its homogenous counterpart by means of suitable substitutions in terms of so-called area and initial potentials.

Problems with Nonhomogeneous Equations and Nonhomogeneous Initial Conditions

9.1 The Time-dependent Area Potential

Let $G = \mathbb{R}^2 \times (0, \infty)$. We define the area potential Uq of density $q \in C_0^\infty(G)$ by

$$(Uq)(X) = \int_G D(x - y, t - \tau) q(y, \tau) \, dy \, d\tau, \quad X \in G,$$

where $D(x, t)$ is the matrix of fundamental solutions computed in §1.2. It is obvious that the Laplace transform $U\hat{q}$ of Uq with respect to the time variable has the form

$$(U\hat{q})(x, p) = \int_{\mathbb{R}^2} \hat{D}(x - y, p) \hat{q}(y, p) \, dy, \quad X \in \mathbb{R}^2.$$

9.1 Lemma. *For any $p \in \mathbb{C}_0$, the operator U can be extended by continuity from $C_0^\infty(\mathbb{R}^2)$ to $H_{-1,p}(\mathbb{R}^2)$. The extended operator is continuous from $H_{-1,p}(\mathbb{R}^2)$ to $H_{1,p}(\mathbb{R}^2)$, and for any $\hat{q} \in H_{-1,p}(\mathbb{R}^2)$, $p \in \mathbb{C}_\kappa$, $\kappa > 0$,*

$$\|U\hat{q}\|_{1,p} \le c|p| \|\hat{q}\|_{-1,p}. \tag{9.1}$$

Proof. First, we establish an estimate for the Fourier transform $\tilde{D}(\xi, p)$ of $\hat{D}(x, p)$ with respect to x. For any $\xi \in \mathbb{R}^2$ and $p \in \mathbb{C}_0$, $\tilde{D}(\xi, p)$ is the solution of the problem

$$Bp^2 \tilde{D}(\xi, p) + (A(\xi)\tilde{D})(\xi, p) = I,$$

where $A(\xi)$ is the matrix

$$\begin{pmatrix} h^2(\lambda + \mu)\xi_1^2 + h^2\mu|\xi|^2 + \mu & h^2(\lambda + \mu)\xi_1\xi_2 & -i\mu\xi_1 \\ h^2(\lambda + \mu)\xi_1\xi_2 & h^2(\lambda + \mu)\xi_2^2 + h^2\mu|\xi|^2 + \mu & -i\mu\xi_2 \\ i\mu\xi_1 & i\mu\xi_2 & \mu|\xi|^2 \end{pmatrix} \tag{9.2}$$

and I is the identity (3×3)-matrix.

Let Ψ be the solution of the equation

$$p^2 B\Psi + A(\xi)\Psi = F \qquad (9.3)$$

for some function F. Multiplying (9.3) by Ψ^*, we obtain

$$p^2 |B^{1/2}\Psi|^2 + \big(A(\xi)\Psi, \Psi\big)_{\mathbb{C}^3} = (F, \Psi)_{\mathbb{C}^3}, \qquad (9.4)$$

where $(F, \Psi)_{\mathbb{C}^3} = F_i \bar{\Psi}_i$ is the standard inner product in $\mathbb{C}^3$. By (9.2),

$$
\begin{aligned}
(A(\xi)\Psi, \Psi)_{\mathbb{C}^3} &= h^2(\lambda + \mu)(\xi_1^2 |\Psi_1|^2 + \xi_2^2 |\Psi_2|^2) \\
&\quad + (h^2 \mu |\xi|^2 + \mu)(|\Psi_1|^2 + |\Psi_2|^2) + \mu |\xi|^2 |\Psi_3|^2 \\
&\quad + 2h^2(\lambda + \mu)\xi_1 \xi_2 \operatorname{Re}(\Psi_1 \bar{\Psi}_2) \\
&\quad + 2\mu \operatorname{Re}\big(i(\xi_1 \Psi_1 + \xi_2 \Psi_2)\bar{\Psi}_3\big) \\
&\geq \mu\big\{\big(\tfrac{1}{2} h^2 |\xi|^2 + 1\big)(|\Psi_1|^2 + |\Psi_2|^2) \\
&\qquad\qquad + (|\xi|^2 - 2h^{-2})|\Psi_3|^2\big\} \\
&\geq \big(k_1(1 + |\xi|^2) - k_2\big)|\Psi|^2 \qquad (9.5)
\end{aligned}
$$

for some positive constants k_1 and k_2.

Separating the real and imaginary parts in (9.4), we arrive at

$$|p|^2 |B^{1/2}\Psi|^2 + \big(A(\xi)\Psi, \Psi\big)_{\mathbb{C}^3} = \sigma^{-1} \operatorname{Re}\{\bar{p}(F, \Psi)_{\mathbb{C}^3}\}, \quad p = \sigma + i\tau. \qquad (9.6)$$

By (9.5) and (9.6), for any $p \in \mathbb{C}_\kappa$, $\kappa > 0$,

$$(1 + |p|^2 + |\xi|^2)|\Psi| \leq c|p|\|F|. \qquad (9.7)$$

If $F = e_j$ is the jth coordinate unit vector and $\Psi = \tilde{D}^{(j)}(\xi, p)$ is the jth column of the matrix $\tilde{D}(\xi, p)$, then, by (9.7),

$$|\tilde{D}^{(j)}(\xi, p)| \leq c|p|(1 + |p|^2 + |\xi|^2)^{-1}, \quad j = 1, 2, 3.$$

Consequently, for any $\hat{q} \in C_0^\infty(\mathbb{R}^2)$,

$$
\begin{aligned}
\|U\hat{q}\|_{1,p}^2 &= \int_{\mathbb{R}^2} (1 + |\xi|^2 + |p|^2)|(\widetilde{U\hat{q}})(\xi, p)|^2 \, d\xi \\
&\leq c|p|^2 \int_{\mathbb{R}^2} (1 + |\xi|^2 + |p|^2)^{-1}|\tilde{q}(\xi, p)|^2 \, d\xi = c|p|^2 \|\hat{q}\|_{-1,p}^2,
\end{aligned}
$$

where $\tilde{q}$ is the Fourier transform of $\hat{q}$. This proves (9.1) and the assertion. $\square$

We now return to the spaces of originals and say that $u \in H_{1,0,\kappa}(G)$ is a weak solution of the Cauchy problem with zero initial data and right-hand side $q \in H_{-1,1,\kappa}(G)$ in the equation of motion if u satisfies

$$\int_0^\infty [a(u,v) - (B^{1/2}\partial_t u, B^{1/2}\partial_t v)_0]\, dt = \int_0^\infty (q,v)_0\, dt \quad \forall v \in C_0^\infty(\bar{G}). \quad (9.8)$$

9.2 Theorem. *For any $q \in H_{-1,1,\kappa}(G)$, $\kappa > 0$, equation (9.8) has a unique solution $u \in H_{1,0,\kappa}(G)$. If $q \in H_{-1,k,\kappa}(G)$, $k \in \mathbb{R}$, then $u \in H_{1,k-1,\kappa}(G)$ and*

$$\|u\|_{1,k-1,\kappa;G} \le c\|q\|_{-1,k,\kappa;G}.$$

Proof. In terms of Laplace transforms, (9.8) assumes the form

$$p^2(B^{1/2}\hat{u}, B^{1/2}\hat{v})_0 + a(\hat{u},\hat{v}) = (\hat{q},\hat{v})_0 \quad \forall \hat{v} \in H_{1,p}(\mathbb{R}^2). \quad (9.9)$$

Clearly,

$$p^2(BU\hat{q})(x,p) + (AU\hat{q})(x,p) = \hat{q}(x,p), \quad x \in \mathbb{R}^2,$$

for any $\hat{q} \in C_0^\infty(\mathbb{R}^2)$; hence, for any $\hat{q} \in H_{-1,p}(\mathbb{R}^2)$, (9.9) admits the solution

$$\hat{u} = U\hat{q} \in H_{1,p}(\mathbb{R}^2).$$

Separating the real and imaginary parts in (9.9) with $\hat{v} = \hat{u}$, that is, in the equality

$$p^2\|B^{1/2}\hat{u}\|_0^2 + a(\hat{u},\hat{u}) = (\hat{q},\hat{u})_0,$$

we arrive at the estimate

$$\|\hat{u}\|_{1,p} \le c|p|\|\hat{q}\|_{-1,p},$$

which shows that (9.9) is uniquely solvable.

We now complete the proof by using (9.9) to establish that the mapping $(U\hat{q})(\cdot, p)$ is holomorphic from $\mathbb{C}_\kappa$ to $H_1(\mathbb{R}^2)$ and that, therefore, $Uq = \mathcal{L}^{-1}U\hat{q}$ exists and is the desired unique solution of (9.8).   □

9.2 The Nonhomogeneous Equation of Motion

With the notation used in Chapter 2, let $q \in H_{-1,1,\kappa}(G^\pm)$. We recall (see §2.3) that $u \in H_{1,0,\kappa}(G^\pm)$ are called weak solutions of the initial-boundary value problems (DD$^\pm$) with a nonhomogeneous equation of motion and homogeneous initial data if they satisfy

$$\int\limits_0^\infty \left[a_\pm(u,v) - (B^{1/2}\partial_t u, B^{1/2}\partial_t v)_{0;S^\pm} \right] dt$$

$$= \int\limits_0^\infty (q,v)_{0;S^\pm}\, dt \quad \forall v \in C_0^\infty(\bar{G}^\pm), \ v^\pm = 0, \qquad (9.10)$$

$$\gamma^\pm u = f,$$

where f and q are prescribed. Theorem 2.3 states conditions for the unique solvability of these problems. In what follows, we show how $(DD^\pm)$ can be reduced to their simpler versions for the corresponding homogeneous equation of motion.

The Laplace transform $\hat{q}(x,p)$ of $q(x,t)$ with respect to t belongs to the space $H_{-1,p}(S^\pm)$ for any $p \in \mathbb{C}_\kappa$. Using extension operators, we construct a function $\hat{Q} \in H_{-1,p}(\mathbb{R}^2)$ such that $\pi^\pm \hat{Q} = \hat{q} \in H_{-1,p}(S^\pm)$ and

$$\|\hat{Q}\|_{-1,p} \le c\|\hat{q}\|_{-1,p;S^\pm}.$$

Obviously, $\mathcal{L}^{-1}\hat{Q} = Q \in H_{-1,1,\kappa}(G)$ and $\pi^\pm Q \in H_{-1,1,\kappa}(G^\pm)$. We now represent the solutions of (9.10) in the form

$$u(X) = w(X) + (UQ)(X), \quad X \in G^\pm,$$

or $u = w + \pi^\pm UQ$. It is clear that the functions $w \in H_{1,0,\kappa}(G^\pm)$ satisfy, respectively, the variational problems

$$\int\limits_0^\infty \left[a_\pm(w,v) - (B^{1/2}\partial_t w, B^{1/2}\partial_t v)_{0;S^\pm} \right] dt$$

$$= \int\limits_0^\infty (q,v)_{0;S^\pm}\, dt$$

$$- \int\limits_0^\infty \left[a_\pm(\pi^\pm UQ, v) - (B^{1/2}\partial_t \pi^\pm UQ, B^{1/2}\partial_t v)_{0;S^\pm} \right] dt$$

$$\forall v \in C_0^\infty(\bar{G}^\pm), \ \text{supp}\, v \subset S^\pm \times [0,\infty),$$

$$\gamma^\pm w = f - \gamma^\pm \pi^\pm UQ.$$

By Theorem 9.2, the right-hand side in the first equality above is zero; consequently, (9.10) reduce to the corresponding problems for the homogeneous equation of motion. If $q \in H_{-1,k,\kappa}(G^\pm)$ and $f \in H_{1/2,k,\kappa}(\Gamma)$, then, by Theorems 9.2 and 2.3, $w \in H_{1,k-1,\kappa}(G^\pm)$ and

$$\|w\|_{1,k-1,\kappa;G^\pm} \le c(\|q\|_{-1,k,\kappa;G^\pm} + \|f\|_{1/2,k,\kappa;\Gamma}).$$

In the case of the initial-boundary value problems (DN$^\pm$) with a nonzero right-hand side $q \in \mathring{H}_{-1,k,\kappa}(G^\pm)$ in the equation of motion and homogeneous initial conditions, we again represent the weak solutions $u \in H_{1,0,\kappa}(G^\pm)$ in the form

$$u(X) = w(X) + (UQ)(X), \quad X \in G^\pm, \tag{9.11}$$

or

$$u = w + \pi^\pm U Q,$$

with Q constructed as above. Arguing in the same way, we then show that representation (9.11) reduces (DN$^\pm$) to the corresponding problems for the homogeneous equation of motion, which were discussed in Chapter 3.

Similar procedures enable us to reduce all the problems considered in the preceding chapters to their counterparts for the homogeneous equation of motion.

9.3 Initial Potentials

In this section, we consider the Cauchy problem in $G = \mathbb{R}^2 \times (0, \infty)$ with the homogeneous equation of motion and nonhomogeneous initial conditions, and represent its solution in the form of some special functions that we call initial single-layer and double-layer potentials.

The classical Cauchy problem in this case consists in finding a function $u \in C^2(G) \cap C^1(\bar{G})$ such that

$$
\begin{aligned}
B(\partial_t^2 u)(X) + (Au)(X) = 0, \quad X \in G, \\
u(x, 0+) = \varphi(x), \quad (\partial_t u)(x, 0+) = \psi(x), \quad x \in \mathbb{R}^2.
\end{aligned}
\tag{9.12}
$$

Let $H_{m,\kappa}(G)$, $m = 0, 1, 2, \ldots$, be the space of three-component vector-valued functions $u(X) = u(x, t)$ defined in G, equipped with the norm

$$\|u\|_{m,\kappa;G}^2 = \sum_{l=0}^{m} \int_0^\infty \int_{\mathbb{R}^2} e^{-2\kappa t} (1 + |\xi|^2)^{m-l} |\partial_t^l \tilde{u}(\xi, t)|^2 \, d\xi \, dt,$$

where $\tilde{u}(\xi, t)$ is the Fourier transform of $u(x, t)$ with respect to x.

We say that $u \in H_{1,\kappa}(G)$ is a weak solution of problem (9.12) if

$$\int_0^\infty \left[a(u, v) - (B^{1/2}\partial_t u, B^{1/2}\partial_t v)_0 \right] dt = (B\psi, \gamma_0 v)_0 \quad \forall v \in C_0^\infty(\bar{G}), \tag{9.13}$$

$$\gamma_0 u = \varphi,$$

where γ_0 is the time trace operator defined by

$$(\gamma_0 u)(x, t) = u(x, 0+), \quad u \in H_{1,\kappa}(G).$$

9.3 Lemma. *The variational problem* (9.13) *has at most one solution.*

This assertion is proved exactly like the uniqueness statement in the proof of Theorem 2.3 with S^+ replaced by $\mathbb{R}^2$.

We now introduce the initial single-layer potential $V_i \psi$ of density $\psi \in L^2(\mathbb{R}^2)$ and the initial double-layer potential $W_i \varphi$ of density $\varphi \in H_1(\mathbb{R}^2)$, respectively, by

$$(V_i\psi)(X) = \int_{\mathbb{R}^2} D(x - y, t) B\psi(y)\, dy, \quad X \in G, \tag{9.14}$$

$$(W_i\varphi)(X) = \int_{\mathbb{R}^2} \partial_t D(x - y, t) B\varphi(y)\, dy, \quad X \in G. \tag{9.15}$$

Going over to Fourier transforms, we see that (9.14) and (9.15) take the form

$$(\widetilde{V_i\psi})(\xi, t) = \tilde{D}(\xi, t) B\tilde{\psi}(\xi), \tag{9.16}$$

$$(\widetilde{W_i\varphi})(\xi, t) = \partial_t \tilde{D}(\xi, t) B\tilde{\varphi}(\xi), \tag{9.17}$$

where $\tilde{D}(\xi, t)$ is the Fourier transform of the matrix of fundamental solutions $D(x, t)$ with respect to x.

In what follows we use the matrix norm

$$\|M\|^2 = \sum_{i,j=1}^{3} |M_{ij}|^2$$

for any (3×3)-matrix $M = (M_{ij})$. As usual, we denote by $M^{(j)}$ the columns of M.

9.4 Lemma. *For any* $(\xi, t) \in \bar{G}$ *and* $l = 0, 1, 2, \ldots,$

$$\|\partial_t^l \tilde{D}(\xi, t)\| \le c_l (1 + t)(1 + |\xi|)^{l-1}. \tag{9.18}$$

Proof. It is obvious that $\tilde{D}(\xi, t) = \theta(t) M(\xi, t)$, where θ is the characteristic function of the interval $(0, \infty)$ and the columns $M^{(j)}$ of M satisfy

$$B(\partial_t^2 M^{(j)})(\xi, t) + A(\xi) M^{(j)}(\xi, t) = 0, \quad (\xi, t) \in G, \tag{9.19}$$
$$M^{(j)}(\xi, 0) = 0, \quad (\partial_t M^{(j)})(\xi, 0) = B^{-1} e_j.$$

By (9.19),

$$\frac{d}{dt}\left\{ |B^{1/2} \partial_t M^{(j)}(\xi, t)|^2 + \left(A(\xi) M^{(j)}(\xi, t), M^{(j)}(\xi, t) \right)_{\mathbb{C}^3} \right\} = 0;$$

hence,

$$|B^{1/2}\partial_t M^{(j)}(\xi,t)|^2 + \big(A(\xi)M^{(j)}(\xi,t), M^{(j)}(\xi,t)\big)_{\mathbb{C}^3} = |B^{-1/2}e_j|^2$$

and

$$|(\partial_t M^{(j)})(\xi,t)| < c. \tag{9.20}$$

By (9.5),

$$(1+|\xi|)^2 |M^{(j)}(\xi,t)|^2 \le c\big\{\big(A(\xi)M^{(j)}(\xi,t), M^{(j)}(\xi,t)\big)_{\mathbb{C}^3}$$
$$+ |M^{(j)}(\xi,t)|^2\big\}$$
$$\le c\{1 + |M^{(j)}(\xi,t)|^2\},$$

and, since

$$|M^{(j)}(\xi,t)| \le \int_0^t |(\partial_\tau M^{(j)})(\xi,\tau)|\, d\tau \le ct,$$

it follows that

$$|M^{(j)}(\xi,t)| \le c(1+t)(1+|\xi|)^{-1}. \tag{9.21}$$

Inequalities (9.20) and (9.21) prove (9.18) for $l = 0,1$.

Suppose that (9.18) holds for $l = 0,1,\ldots,k$, $k \ge 1$. Differentiating the equation in (9.19) with respect to t, we find that

$$(\partial_t^{k+1} M^{(j)})(\xi,t) = -B^{-1}A(\xi)(\partial_t^{k-1} M^{(j)}), \quad (\xi,t) \in G;$$

therefore,

$$|(\partial_t^{k+1} M^{(j)})(\xi,t)| \le c(1+|\xi|)^2(1+t)(1+|\xi|)^{k-2} = c(1+t)(1+|\xi|)^k,$$

which implies that (9.18) holds for all $l = 0,1,2,\ldots$. $\square$

9.5 Theorem. *For any $\psi \in L^2(\mathbb{R}^2)$, $\varphi \in H_1(\mathbb{R}^2)$, and $\kappa > 0$, the function*

$$u = V_i\psi + W_i\varphi$$

is the (unique) weak solution of the Cauchy problem (9.13). If $\psi \in H_k(\mathbb{R}^2)$ and $\varphi \in H_{k+1}(\mathbb{R}^2)$, $k = 0,1,2,\ldots$, then $u \in H_{k+1,\kappa}(G)$ and

$$\|u\|_{k+1,\kappa;G} \le c(\|\psi\|_k + \|\varphi\|_{k+1}). \tag{9.22}$$

Proof. By (9.16) and (9.17),

$$\tilde{u}(\xi,t) = \tilde{D}(\xi,t)B\tilde{\psi}(\xi) + \partial_t\tilde{D}(\xi,t)B\tilde{\varphi}(\xi);$$

therefore,

$$\|u\|_{k+1,\kappa;G}^2$$

$$= \sum_{l=0}^{k+1} \int_0^\infty \int_{\mathbb{R}^2} e^{-2\kappa t}(1+|\xi|^2)^{k+1-l}$$

$$\times |\partial_t^l \tilde{D}(\xi,t)B\tilde{\psi}(\xi) + \partial_t^{l+1}\tilde{D}(\xi,t)B\tilde{\varphi}(\xi)|^2 \, d\xi$$

$$\leq c\sum_{l=0}^{k+1} \int_0^\infty \int_{\mathbb{R}^2} e^{-2\kappa t}(1+|\xi|^2)^{k+1-l}\|\partial_t^l \tilde{D}(\xi,t)\|^2|\tilde{\psi}(\xi)|^2 \, d\xi \, dt$$

$$+ c\sum_{l=0}^{k+1} \int_0^\infty \int_{\mathbb{R}^2} e^{-2\kappa t}(1+|\xi|^2)^{k+1-l}\|\partial_t^{l+1}\tilde{D}(\xi,t)\|^2|\tilde{\varphi}(\xi)|^2 \, d\xi \, dt,$$

so, by (9.18),

$$\|u\|_{k+1,\kappa;G}^2$$

$$\leq c\int_0^\infty \int_{\mathbb{R}^2} e^{-2\kappa t}(1+t)^2(1+|\xi|^2)^{k+1-l}$$

$$\times \left\{(1+|\xi|^2)^{l-1}|\tilde{\psi}(\xi)|^2 + (1+|\xi|^2)^l|\tilde{\varphi}(\xi)|^2\right\} d\xi \, dt$$

$$\leq c(\|\psi\|_k^2 + \|\varphi\|_{k+1}^2),$$

which proves (9.22).

We now show that $u(x,t) \to \varphi(x)$ in $H_{1/2}(\mathbb{R}^2)$ as $t \to 0+$. We have

$$\|u(\cdot,t) - \varphi\|_{1/2}^2 = \int_{\mathbb{R}^2} (1+|\xi|)|\tilde{u}(\xi,t) - \tilde{\varphi}(\xi)|^2 \, d\xi$$

$$\leq c\left\{ \int_{\mathbb{R}^2} (1+|\xi|)\|\tilde{D}(\xi,t)\|^2|\tilde{\psi}(\xi)|^2 \, d\xi \right.$$

$$\left. + \int_{\mathbb{R}^2} (1+|\xi|)\|\partial_t\tilde{D}(\xi,t)B - I\|^2|\tilde{\varphi}(\xi)|^2 \, d\xi \right\}.$$

Since $\tilde{D}(\xi,t) \to 0$ and $(\partial_t\tilde{D})(\xi,t) \to B^{-1}$, as $t \to 0+$, for any $\xi \in \mathbb{R}^2$, from Lebesgue's dominated convergence theorem and (9.18) it follows that

$$\lim_{t\to 0+} \|u(\cdot,t) - \varphi\|_{1/2} = 0.$$

To complete the proof, it remains to verify that u satisfies the variational equation in (9.13). Let φ, $\psi \in C_0^\infty(\mathbb{R}^2)$. By (9.22), u is a smooth function. Going over to Fourier transforms with respect to x in (9.13), integrating by parts with respect to t and taking the initial data in (9.19) into account, we conclude that u indeed satisfies the relevant variational equation. The proof of the assertion now follows from (9.22) and the fact that $C_0^\infty(\mathbb{R}^2)$ is dense in $H_k(\mathbb{R}^2)$. $\qquad\qquad\qquad\qquad\qquad\qquad\qquad\qquad\qquad\qquad\qquad\qquad$ □

9.6 Remark. The above analysis suggests the following solution procedure for problems (DD$^\pm$) with boundary data f and nonhomogeneous initial data φ and ψ. If $\varphi \in H_{k+1}(S^\pm)$ and $\psi \in H_k(S^\pm)$ for some $k = 0, 1, 2, \ldots,$ then we use extension operators to construct

$$\Phi \in H_{k+1}(\mathbb{R}^2), \quad \pi^\pm \Phi = \varphi \quad \text{and} \quad \Psi \in H_k(\mathbb{R}^2), \quad \pi^\pm \Psi = \psi,$$

and represent the solutions u in the form

$$u(X) = u_0(X) + w(X), \quad X \in G^\pm,$$

where

$$u_0 = \pi^\pm V_i \Psi + \pi^\pm W_i \Phi.$$

Clearly, w must satisfy the same equation of motion as u but with homogeneous initial data. The boundary condition for w takes the form

$$\gamma^\pm w = f - \gamma^\pm u_0.$$

If

$$f - \gamma^\pm u_0 \in H_{1/2,l,\kappa}(\Gamma), \tag{9.23}$$

then our problems have been reduced to those considered in §2.3. It turns out that (9.23) is the compatibility condition for the initial and boundary data. A more detailed discussion of such compatibility is outside the scope of this book. The interested reader is referred, for example, to [13].

Analogous techniques can also be set up for the other problems in the preceding chapters, to reduce their general form with nonhomogeneous initial conditions to their corresponding homogeneous counterparts.

A

The Fourier and Laplace Transforms of Distributions

Below we introduce the Fourier and Laplace transformations for test functions and distributions and list (without proof) some of their most important properties.

A.1 Definition. A complex-valued function φ defined on $\mathbb{R}^m$ is called a test function (of rapid descent) if $\varphi \in C^\infty(\mathbb{R}^m)$ and for any $k \in \mathbb{Z}_+$,

$$\langle \varphi \rangle_k = \sup_{x \in \mathbb{R}^m} (1 + |x|)^k \sum_{|\alpha| \le k} |\partial^\alpha \varphi(x)| < \infty,$$

where $|x| = (x_1^2 + \cdots + x_m^2)^{1/2}$. The space of all such functions is denoted by $\mathcal{S}(\mathbb{R}^m)$.

A.2 Definition. The Fourier transform $\tilde{\varphi}$ of a test function (of rapid descent) $\varphi \in \mathcal{S}(\mathbb{R}^m)$ is defined by

$$\tilde{\varphi}(\xi) = (\mathcal{F}\varphi)(\xi) = \int_{\mathbb{R}^m} e^{i(x,\xi)} \varphi(x) \, dx. \tag{A.1}$$

A.3 Theorem. *The operator $\mathcal{F}$ is a homeomorphism from $\mathcal{S}(\mathbb{R}^2)$ to $\mathcal{S}(\mathbb{R}^2)$, and its inverse $\mathcal{F}^{-1}$ acts according to the formula*

$$\varphi(x) = (\mathcal{F}^{-1}\tilde{\varphi})(x) = (2\pi)^{-m} \int_{\mathbb{R}^m} e^{-i(x,\xi)} \tilde{\varphi}(\xi) \, d\xi, \quad \tilde{\varphi} \in \mathcal{S}(\mathbb{R}^m).$$

A.4 Remark. The continuity of $\mathcal{F}$ from $\mathcal{S}(\mathbb{R}^m)$ to $\mathcal{S}(\mathbb{R}^m)$ means that for any $k \in \mathbb{Z}_+$, there is $l(k)$ such that

$$\langle \mathcal{F}\varphi \rangle_k \le c_k \langle \varphi \rangle_{l(k)} \quad \forall \varphi \in \mathcal{S}(\mathbb{R}^m).$$

A.5 Theorem. (i) *For any* $\varphi, \psi \in S(\mathbb{R}^m)$, *there holds Parseval's equality*

$$\int_{\mathbb{R}^m} \tilde{\varphi}(\xi)\overline{\tilde{\psi}(\xi)}\, d\xi = (2\pi)^m \int_{\mathbb{R}^m} \varphi(x)\overline{\psi(x)}\, dx. \qquad (A.2)$$

(ii) $\mathcal{F}(\varphi * \psi)(\xi) = \tilde{\varphi}(\xi)\tilde{\psi}(\xi)$ *for any* $\varphi, \psi \in S(\mathbb{R}^m)$.
(iii) *For any* $\varphi \in S(\mathbb{R}^2)$ *and any multiindex* α,

$$\mathcal{F}(\partial^\alpha \varphi)(\xi) = (-i)^{|\alpha|} \xi^\alpha \tilde{\varphi}(\xi),$$
$$\mathcal{F}(x^\alpha \varphi)(\xi) = (-i)^{|\alpha|} \partial^\alpha \tilde{\varphi}(\xi).$$

A.6 Example. If $m = 1$ and $\varphi(x) = e^{-a^2 x^2}$, $a > 0$, then

$$(\mathcal{F} e^{-a^2 x^2})(\xi) = \int_{\mathbb{R}} e^{ix\xi - a^2 x^2}\, dx$$

$$= \frac{1}{a} e^{-\xi^2/(4a^2)} \int_{\mathbb{R}} e^{-(\sigma + i\xi/(2a))^2}\, d\sigma$$

$$= \frac{1}{a} e^{-\xi^2/(4a^2)} \int_{\mathrm{Im}\, \zeta = \xi/(2a)} e^{-\zeta^2}\, d\zeta.$$

Using Cauchy's theorem, we can easily check that

$$\int_{\mathrm{Im}\, \zeta = \xi/(2a)} e^{-\zeta^2}\, d\zeta = \int_{\mathbb{R}} e^{-\zeta^2}\, d\zeta = \sqrt{\pi};$$

therefore,

$$(\mathcal{F} e^{-a^2 x^2})(\xi) = \frac{\sqrt{\pi}}{a} e^{-\xi^2/(4a^2)}.$$

A.7 Remark. Since $S(\mathbb{R}^m)$ is dense in $L^1(\mathbb{R}^m)$ and $L^2(\mathbb{R}^m)$, the Fourier transformation can be extended by continuity to the latter spaces. Thus, if $f \in L^1(\mathbb{R}^m)$, then $\mathcal{F}f$ is simply defined by (A.1) and is continuous and bounded in $\mathbb{R}^m$.

If $f \in L^2(\mathbb{R}^m)$, then we approximate f by a sequence $\{f\}_{n=1}^\infty$ in $L^2(\mathbb{R}^m) \cap L^1(\mathbb{R}^m)$, for example, the sequence of truncations

$$f_n(x) = \begin{cases} f(x), & |x| < n, \\ 0, & |x| > n. \end{cases}$$

It is not difficult to show that $f_n \to f$ and $\tilde{f}_n \to \tilde{f} \in L^2(\mathbb{R}^m)$. Then we again define the Fourier transform of f by (A.1), understanding that equality in the above sense.

A.8 Theorem. *If $f, g \in L^2(\mathbb{R}^m)$, then*

(i) *the inverse transformation $\mathcal{F}^{-1}$ acts according to the formula*

$$f(x) = (\mathcal{F}^{-1}\tilde{f})(x) = (2\pi)^{-m} \int_{\mathbb{R}^m} e^{-i(x,\xi)} \tilde{f}(\xi) \, d\xi.$$

(ii) *Parseval's equality (A.2) holds.*

A.9 Definition. The (generalized) Fourier transform

$$\tilde{f} = \mathcal{F}f$$

of a tempered distribution $f \in \mathcal{S}'(\mathbb{R}^m)$ (an element of the dual of $\mathcal{S}(\mathbb{R}^m)$) is defined by

$$(\tilde{f}, \tilde{\varphi}) = (2\pi)^m (f, \varphi) \quad \forall \tilde{\varphi} \in \mathcal{S}(\mathbb{R}^m),$$

where

$$\varphi = \mathcal{F}^{-1}\tilde{\varphi}$$

and $(\cdot, \cdot)$ is the duality generated by the inner product in $L^2(\mathbb{R}^m)$.

A.10 Theorem. (i) *The distributional operator $\mathcal{F}$ is a homeomorphism from $\mathcal{S}'(\mathbb{R}^m)$ to $\mathcal{S}'(\mathbb{R}^m)$.*

(ii) *For any $f \in \mathcal{S}'(\mathbb{R}^m)$ and any $\varphi \in \mathcal{S}(\mathbb{R}^m)$,*

$$\mathcal{F}(f * \varphi) = \tilde{\varphi}\tilde{f}.$$

(iii) *For any $f \in \mathcal{S}'(\mathbb{R}^m)$ and any multiindex α,*

$$\mathcal{F}(\partial^\alpha f) = (-i)^{|\alpha|} \xi^\alpha \tilde{f},$$
$$\mathcal{F}(x^\alpha f) = (-i)^{|\alpha|} \partial^\alpha \tilde{f}.$$

A.11 Examples. (i) If δ is the Dirac delta, then, by Theorem A.3(iii),

$$\mathcal{F}(\partial^\alpha \delta) = (-i)^{|\alpha|} \xi^\alpha \tilde{\delta}.$$

Since for any $\tilde{\varphi} \in \mathcal{S}(\mathbb{R}^m)$,

$$(\tilde{\delta}, \tilde{\varphi}) = (2\pi)^m (\delta, \varphi) = (2\pi)^m \overline{\varphi(0)} = \int_{\mathbb{R}^m} \overline{\tilde{\varphi}(\xi)} \, d\xi = (1, \tilde{\varphi}),$$

it follows that

$$\mathcal{F}(\partial^\alpha \delta) = (-i)^{|\alpha|} \xi^\alpha.$$

(ii) For the one-dimensional distribution $\widetilde{x^{-1}}$ generated by the function x^{-1} (in the sense of principal value) we have

$$(\mathcal{F}\widetilde{x^{-1}}, \tilde{\varphi}) = 2\pi(x^{-1}, \varphi) = 2\pi \lim_{\substack{\varepsilon \to 0 \\ R \to \infty}} \int_{\varepsilon < |x| < R} \frac{1}{x} \overline{\varphi(x)} \, dx$$

$$= \lim_{\substack{\varepsilon \to 0 \\ R \to \infty}} \int_{\mathbb{R}} \left(\int_{\varepsilon < |x| < R} \frac{1}{x} e^{ix\xi} \, dx \right) \overline{\tilde{\varphi}(\xi)} \, d\xi$$

$$= 2i \lim_{\substack{\varepsilon \to 0 \\ R \to \infty}} \int_{\mathbb{R}} \left(\int_{\varepsilon}^{R} \frac{1}{x} \sin(x\xi) \, dx \right) \overline{\tilde{\varphi}(\xi)} \, d\xi.$$

Since

$$\int_{0}^{\infty} \frac{1}{x} \sin(x\xi) \, dx = \tfrac{1}{2} \pi \operatorname{sgn} \xi,$$

we finally obtain

$$\mathcal{F}\widetilde{x^{-1}} = i\pi \operatorname{sgn} \xi.$$

We now turn our attention to the Laplace transformation for one-dimensional distributions. Let $g \in L^2(\mathbb{R})$, $\operatorname{supp} g \in [0; \infty)$, and suppose that there is $a \in \mathbb{R}$ such that $e^{-\xi t}g(t)$ belongs to $L^2(0; \infty)$ for $\xi > a$ and $e^{-\xi t}g(t)$ does not belong to $L^1(0; \infty)$ for $\xi < a$. If $a < a' < \xi$, then the equality

$$e^{-\xi t}g(t) = e^{-(\xi - a')t}e^{-a't}g(t)$$

implies that $e^{-\xi t}g(t)$ belongs to $L^1(0; \infty)$ for $\xi > a$. In this case, the Laplace transform $(\mathcal{L}g)(p)$, $p = \xi + i\eta$, $\xi > a$, exists and is defined by

$$G(p) = (\mathcal{L}g)(p) = \int_{0}^{\infty} e^{-pt}g(t) \, dt.$$

The number a is called the Dedekind abscissa of absolute convergence of the Laplace transform.

A.12 Theorem. *The function $G(p) = G(\xi + i\eta)$ is holomorphic in the half-space $\operatorname{Re} p = \xi > a$.*

A.13 Remark. We may write the Laplace transform G of g in the form

$$G(p) = G(\xi + i\eta) = \int_{0}^{\infty} e^{-it\eta}e^{-\xi t}g(t) \, dt = \left(F(e^{-\xi t}g)\right)(-\eta). \tag{A.3}$$

By Parseval's equality,

$$2\pi \int_0^\infty e^{-2\xi t} |g(t)|^2 \, dt = \int_{\mathbb{R}} |G(\xi + i\eta)|^2 \, d\eta;$$

hence,

$$\int_0^\infty e^{-2at} |g(t)|^2 \, dt = \sup_{\xi > a} \int_0^\infty e^{-2\xi t} |g(t)|^2 \, dt = \frac{1}{2\pi} \sup_{\xi > a} \int_{\mathbb{R}} |G(\xi + i\eta)|^2 \, d\eta.$$

From (A.3) it now follows that if $\xi > a$, then

$$e^{-\xi t} g(t) = \frac{1}{2\pi} \int_{\mathbb{R}} e^{-i\eta t} G(\xi - i\eta) \, d\eta = \frac{1}{2\pi} \int_{\mathbb{R}} e^{i\eta t} G(\xi + i\eta) \, d\eta,$$

so the inverse Laplace transformation is defined by the formula

$$g(t) = (\mathcal{L}^{-1} G)(t) = \frac{1}{2\pi i} \int_{\operatorname{Re} p = \xi} e^{pt} G(p) \, dp.$$

Clearly, the convergence of these integrals is understood in the L^2-sense.

Let g and h be functions of the above class with abscissas of absolute convergence a and b, respectively. The convolution $u = g * h$ of these functions is defined by

$$u(t) = (g * h)(t) = \int_0^t g(t - \tau) h(\tau) \, d\tau.$$

A.14 Theorem. *The Laplace transform $U(p) = (\mathcal{L}u)(p)$ exists, is holomorphic for $\xi > \max\{a, b\}$, and*

$$U(p) = G(p) H(p),$$

where G and H are the Laplace transforms of f and h, respectively.

Next, we define the Laplace transform of a distribution. We restrict the space of test functions (of rapid descent) for $m = 1$ to the subspace $\mathcal{D}(\mathbb{R}) = C_0^\infty(\mathbb{R})$ of all infinitely differentiable functions with compact support in $\mathbb{R}$, and denote by $\mathcal{D}'_+(\mathbb{R})$ the set of distributions in $\mathcal{D}'(\mathbb{R})$ (the dual of $\mathcal{D}(\mathbb{R})$) with support in $[0; \infty)$.

Let $g \in \mathcal{D}'_+(\mathbb{R})$, and let $a \in \mathbb{R}$ be such that $e^{-\xi t} g(t)$ belongs to $\mathcal{S}'(\mathbb{R})$ for $\xi > a$ and $e^{-\xi t} g(t)$ does not belong to $\mathcal{S}'(\mathbb{R})$ for $\xi < a$. The number a in this case is called the abscissa of convergence of the Laplace transform of the distribution g. It is easy to show that if $\xi > a$, then $e^{-\xi t}g(t)$ belongs to $\mathcal{S}'(\mathbb{R})$.

To define the distributional (generalized) Laplace transformation, we first introduce the distributional equivalent of the change of variable for functions, which takes us from $f(x)$ to $f(-x)$. Specifically, for any $f \in \mathcal{S}'(\mathbb{R})$, we define $f_{-x} \in \mathcal{S}'(\mathbb{R})$ by

$$\left(f_{-x}, \varphi(x)\right) = \left(f, \varphi(-x)\right) \quad \forall \varphi \in \mathcal{D}(\mathbb{R}).$$

A.15 Definition. If $g \in \mathcal{D}'_+(\mathbb{R})$ is a distribution with abscissa of convergence $a \in \mathbb{R}$, then its (generalized) Laplace transform $G(p) = (\mathcal{L}g)(p)$ is defined for $\xi = \operatorname{Re} p > a$ by

$$G(p) = (\mathcal{L}g)(p) = \left(F(e^{-\xi t}g)\right)_{-\eta}(p), \quad p = \xi + i\eta.$$

A.16 Theorem. *If g and h are elements of $\mathcal{D}'_+(\mathbb{R})$ with abscissas of convergence a and b and Laplace transforms G and H, respectively, then*
  (i) *$G(p)$, $p = \xi + i\eta$, is holomorphic for $\xi > a$;*
  (ii) *$(\mathcal{L}(\partial_t^l g))(p) = p^l G(p)$;*
  (iii) *$u = g * h$ exists and*

$$U(p) = (\mathcal{L}u)(p) = G(p)H(p) \quad \text{for } \xi = \operatorname{Re} p > \max\{a, b\}.$$

A.17 Theorem. *Let $g \in \mathcal{D}'_+(\mathbb{R})$ be a distribution with abscissa of convergence a and Laplace transform G.*
  (i) *For any $\varepsilon > 0$, there are positive numbers c and l such that*

$$|G(p)| \leq c(1 + |p|)^l \quad \text{for } \operatorname{Re} p = \xi \geq a + \varepsilon; \tag{A.4}$$

that is, $G(p)$ grows at infinity no faster than a polynomial.
  (ii) *If $G(p)$ is holomorphic for $\xi > a$ and satisfies (A.4), then there is a unique distribution $g \in \mathcal{D}'_+(\mathbb{R})$ such that*

$$(\mathcal{L}g)(p) = G(p) \quad \text{for } \xi > a.$$

A.18 Example. It is easy to verify that the Laplace transform of the one-dimensional Dirac delta is $(\mathcal{L}\delta)(p) = 1$.

References

1. Abramowitz, M., Stegun, I.: Handbook of Mathematical Functions. Dover (1964)
2. Agranovich, M.S., Vishik, M.I.: Elliptic problems with a parameter and parabolic problems of a general form. Russian Math. Surveys, 19:3, 53–161 (1964)
3. Akhiezer, N.I., Glazman, I.M.: Theory of Linear Operators in Hilbert Space, vol. 1. F. Ungar (1961–63)
4. Chudinovich, I., Constanda, C.: Non-stationary integral equations for elastic plates. C.R. Acad. Sci. Paris Sér. I, 329, 1115–1120 (1999)
5. Chudinovich, I., Constanda, C.: Solvability of initial-boundary value problems in bending of plates. J. Appl. Math. Phys., 51, 449–466 (2000)
6. Chudinovich, I., Constanda, C.: The Cauchy problem in the theory of plates with transverse shear deformation. Math. Models Methods Appl. Sci., 10, 463–477 (2000)
7. Chudinovich, I., Constanda, C.: Variational and Potential Methods in the Theory of Bending of Plates with Transverse Shear Deformation. Chapman & Hall/CRC (2000)
8. Chudinovich, I., Constanda, C.: Dynamic transmission problems for plates. J. Appl. Math. Phys. (ZAMP), 53, 1060–1074 (2002)
9. Constanda, C.: A Mathematical Analysis of Bending of Plates with Transverse Shear Deformation. Longman/Wiley (1990)
10. Constanda, C.: Direct and Indirect Boundary Integral Equation Methods. Chapman & Hall/CRC (1999)
11. Doetsch, G.: Anleitung zum praktischen Gebrauch der Laplace-Transformation und der Z-Transformation. R. Oldenbourg (1967)
12. Eskin, G.I.: Boundary Value Problems for Elliptic Pseudodifferential Equations. Amer. Math. Soc. (1981)
13. Lions, J.-L., Magenes, E.: Nonhomogeneous Boundary Value Problems and Applications, vol. 1. Springer-Verlag (1972)
14. Love, A.E.H.: A Treatise on the Mathematical Theory of Elasticity. Cambridge University Press (1959)
15. Lubich, Ch.: On the multistep time discretization of linear initial-boundary value problems and their boundary integral equations. Numer. Math., 67, 364–389 (1994)
16. Mizohata, S.: The Theory of Partial Differential Equations. Cambridge University Press (1973)
17. Sokolnikoff, I.S.: Mathematical Theory of Elasticity. McGraw-Hill (1956)
18. Yosida, K.: Functional Analysis. Springer-Verlag (1966)

Index